30117 010155766

D1643840

THE UNSEEN UNIVERSE

Photographs from the Scanning Electron Microscope

C. P. Gilmore

Schocken Books · New York

First SCHOCKEN PAPERBACK edition 1974

Published by arrangement with the New York Graphic Society

Manufactured in the United States of America

Library of Congress Cataloging in Publication Data

Gilmore, Clarence Percy, 1926-
The unseen universe.

First ed. published under title: The scanning electron microscope.
1. Scanning electron microscope—Pictorial works.
I. Title.
[QH212.S3G55 1974] 502'.8 74-10141

CONTENTS

introduction: beauty in the unseen universe 6

CHAPTER 1 • small creatures 11

Carniverous Plant 12
Flies 14
Ants 20
Gnats 24
Mosquitoes 26
Bees 29
Butterfly Wings 30
Mites 32
The Case of the Clinging Gekko 36

CHAPTER 2 • the world of the sea 39

Structures from the Sea 40
Fabulous Fossils 44
Diatoms 46
Mother-of-Pearl 48
Sea Urchin 50

CHAPTER 3 • around the house 53

Common Objects 54
Meat and Potatoes 58
Fibers 60
Carpets 66

CHAPTER 4 • function follows form 69

Sensors: Sight and the Eye 71
Hearing and the Ear 73
Taste and the Tongue 76
Chromosomes 80
Cilia 82
Digestion 83
Skin and Hair 85

CHAPTER 5 • the physical universe 93

Dust from Another World 94
Minerals 96
Pollution 105

CHAPTER 6 • specialized cells 107

Neurons 108
Cell Division 110
Blood Cells 112
Sperm 118

CHAPTER 7 • minute life: animal, vegetable, and in-between 123

Bacteria Colony 124
How Do Antibiotics Kill Bacteria? 129
Effect of DDT on Bird Eggs 130
Slime Mold 132
Pollen 138
Soil Protozoa 156

INTRODUCTION

Beauty in the Unseen Universe

There is a serene beauty in the order of nature that cannot be accounted for by chance. The economical and elegant solutions developed by living creatures under evolutionary pressures are clearly functional as well as aesthetically pleasing.

This principle, an article of faith among many scientists despite the fact that it is probably not provable, has been confirmed again—this time in the sub-microscope world—by a startling new instrument called the scanning electron microscope. It opens a window into a previously unseen world.

The photographs in this book were almost all taken by scientists for scientific purposes. Many of them, as the captions detail, have provided new insight and understanding in a wide variety of scientific areas.

Yet many of them also have a haunting beauty that cannot be without purpose, although no one can say what the purpose is.

The pictures that follow are unique; nothing like them has ever been seen before. If you have leafed through this volume, this remark may strike you as curious. The photos, especially those of insects and other recognizable objects, seem realistic and lifelike, as though they had been taken with an ordinary camera. Yet they are far from ordinary.

The normal view through a microscope—either the optical kind you may have used in Biology I or the complex electron microscope used by research organizations—is usually a meaningless blur of incomprehensible patterns, shadows, and forms, meaningful only to the investigator who knows what he is looking for.

The *scanning* electron microscope, on the other hand, produces realistic looking photographs that are giving scientists a view of a hitherto unseen realm. Now, for the first time, investigators can see details and features of microscopic objects previously difficult or impossible to obtain.

The new instrument is producing striking results in many fields: biology, zoology, electronics, geology, and metallurgy, to name a few. Its acceptance by the scientific community, consequently, has been striking. The first commercial units became available only in the late 1960's; already thousands are in use in laboratories around the world, despite the fact that they cost in the neighborhood of $100,000. As one researcher put it, "This is probably the fastest growth of the use of any scientific instrument in history."

The scanning electron microscope works in a totally different way from either of the other basic types of microscopes, the light microscope and the *transmission* electron microscope.

The light microscope—the kind laboratory technicians and college students usually squint through—magnifies an image because its lens refracts light in a useful way. This extraordinarily important instrument brought about a total revolution in our understanding of life. When early investigators saw unbelievably large numbers of previously unsuspected creatures swimming in a single drop of stagnant water, they caught the first glimpses of an entirely new world, the full depths of which have not yet been plumbed.

But despite its central role in the biological revolution of the last century, the light microscope has its limitations. The German physicist Ernst Abbe concluded a century ago that the light microscope had a maximum theoretical resolving power of about 1/125,000th of an inch; that is, it would never be able to distinguish between two objects separated by less distance than that or see objects smaller than that size.

Abbe based his calculations on the fact that light in many ways resembles radio waves. And visible light falls within a certain range of wavelengths. He calculated that it would never be possible to see anything much smaller than the wavelength of the light being used to view the object.

Unfortunately, there were many objects scientists wanted to see that are smaller than 1/125,000th of an inch, or about 2,000 angstroms, to use the scientific term. So investigators began casting around for some way to get beyond the light barrier.

The first evidence that it might be possible came in 1924. The French physicist Louis deBroglie discovered what is now known as the dual nature of electrons. While these ultra-small components of atoms behave in some ways like particles, in others they appear to be a wave phenomenon, like light or radio signals. It was also clear that if a microscope could make use of this wave-like nature of the electron, it would bring about an enormous advance in microscopy. The wavelength of the electron was thousands of times shorter than that of visible light. So a microscope that illuminated its subject with a beam of electrons rather than a beam of light would have the theoretical potential of resolving objects thousands of times smaller than the minimum size resolved by the light microscope. The first electron microscopes designed to test this theory were built in Germany in the early 1930's.

In principle, the electron microscope is identical to the light microscope. It directs a beam of electrons through a sample; the beam carrying the image passes through a series of lenses that magnify it.

The physical equipment is of course quite different. Magnetic coils designed to refract the beam of electrons form a magnetic "lens" that performs the same function for electrons that the optical lens does for light. The electron beam, bearing the magnified image of the specimen, eventually hits either a fluorescent screen or a piece of photographic film, so the image can be viewed or recorded.

The electron microscope proved as powerful in practice as it did in theory. Where the optical microscope can resolve 2,000 angstroms, the modern electron instrument does a thousand times better. It reveals minute particles such as viruses, no more than five angstroms across. A careful operator can sometimes resolve two or three angstroms.

The transmission electron microscope brought about a quantum leap in the life and physical sciences. But there was still a troublesome limitation. Both the light and the conventional electron microscope are *transmission* instruments. The illuminating beam—light in the one case, electrons in the other—shines *through* a thin specimen. The resultant picture can be thought of as something like an X-ray. Like an X-ray, it can contain a great deal of useful information. But also like an X-ray, it bears little resemblance to the object's actual appearance.

The *scanning* electron microscope, for the first time, supplies a real-life view of the microscopic world. It does so because it photographs the *surface* of solid objects in exactly the same way an ordinary camera photographs the surface of its subject. This accounts for the startlingly realistic look of the scanning electron microscope photographs.

The scanning electron microscope not only presents a new view of nature; it also saves a tremendous amount of time. Previously, when an investigator wanted to get a complete view of a specimen, he had to take it carefully through a sequence of chemical processes to preserve it, then slice it into many microscopically thin sections with a device called a microtome. Each of these sections was painstakingly placed on a slide, put in the microscope, and photographed. Finally, an artist could take all of the cross-section views and reconstruct a three-dimensional approximation.

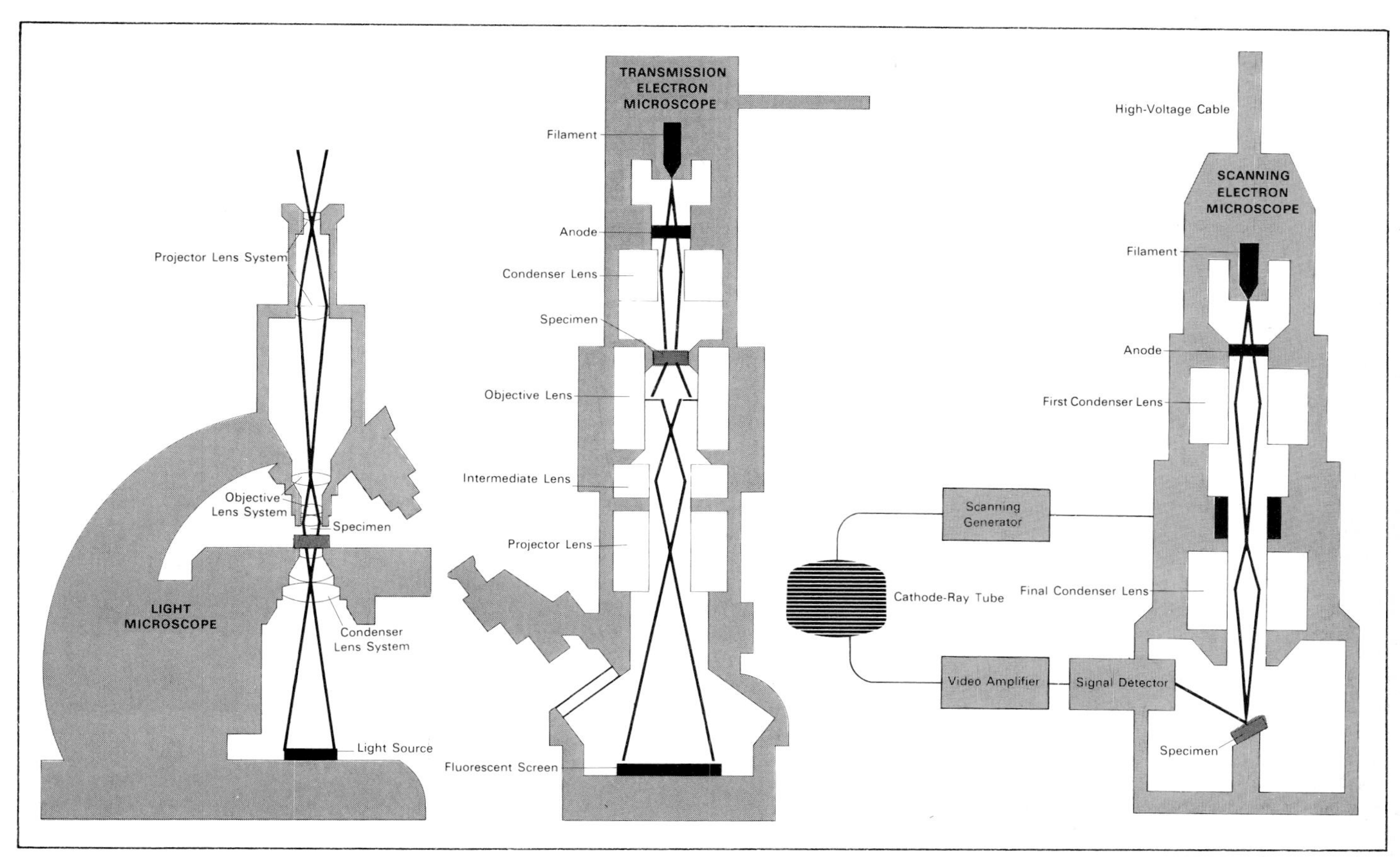

THE LIGHT MICROSCOPE magnifies images by sending a beam of light through an object, then manipulating the light beam with a series of optical lenses usually made of glass. The condenser lens focuses the light beam on the specimen; the objective lenses magnify the beam, which now contains the image; the projector lenses magnify the image again and focus it on the eye. The light microscope can magnify objects usefully about 2,500 times. Actually it would be simple to make a light microscope that magnified 10 or 100 or 1000 times more, but it would accomplish nothing—no more detail would appear in the image. Magnification, in fact, is not a good indicator of a microscope's capability, and professionals judge an instrument's *resolving* power instead. The best optical microscopes can resolve or reveal details as small as 2000 angstroms—that's about 1/125,000th of an inch. Physical laws having to do with the fundamental nature of light bar any greater resolution from visible light microscopes.

THE TRANSMISSION ELECTRON MICROSCOPE is similar in principle to the light microscope. Its filament generates an electron beam that passes through a series of lens systems and the specimen to be viewed. The three basic lenses perform the same functions they do in the light microscope, but they focus and refract a beam of electrons instead of a light beam. Thus they are made in the form of a series of magnetic coils, which affect an electron beam in much the same way glass lenses affect light. Since the eye cannot see electrons, the final projection lens focuses the image-carrying beam on a fluorescent screen where it can be seen, or on a photographic plate where the image can be permanently recorded. Since the electron beam is not subject to the physical laws that limit the resolving power of light, the electron microscope can do a thousand times better; in the hands of a skilled operator it can reveal ultra-small objects such as viruses no more than 2 or 3 angstroms across.

THE SCANNING ELECTRON MICROSCOPE works in an entirely different way. Instead of shining a beam through the specimen, it forms the electrons into an exceedingly fine beam perhaps 200 angstroms across (less than a millionth of an inch). Then it sweeps the beam across the specimen in exactly the same way a single spot of light sweeps across the face of a television screen to build up an entire picture. As the beam sweeps across in a series of step-like passes, the electrons strike the specimen over its entire surface. At each point, the impinging electrons knock loose showers of electrons that are part of the atoms of which the specimen is made. The number of these "secondary" electrons ejected from any given point is determined by the composition of the specimen at that point and the angle at which the beam strikes. For example, the beam generates many more secondary electrons when it slams into the target broadside than when it hits a glancing blow. The stream of secondary electrons, which is constantly varying according to the nature of the surface the beam is scanning, is picked up by a signal detector, amplified, and applied to the face of a TV tube whose scan is in synchronization with the electron scan inside the microscope. Thus, line by line, the picture tube builds up a perfect replica of the object being scanned by the beam. The photos in this volume are simply photographs of the picture tube built into the instrument. They look much sharper than ordinary TV pictures because the system uses approximately twice as many scanning lines as commercial TV—1000 as opposed to 525. Thus resolution is good enough so that the lines in many pictures cannot be seen, although a careful observer will be able to see scanning lines in some photos.

Such a process can take weeks or months to produce a single composite picture. Therefore it is impractical for an investigator to view hundreds or thousands of objects to study their similarities and differences. Now, with the scanning electron microscope, a new sample can be viewed and photographed every few minutes, and large-scale surveys suddenly become not only practical, but relatively simple.

Sample preparation for the scanning electron microscope is quick and easy. First, the specimen must be fixed—chemically treated so that it will retain its original size and shape. Then it must be metal plated to make it conductive. Otherwise, the electron beam used to scan the specimen can deposit a static charge that distorts the picture. The plating is done by a simple process called gold evaporation. The specimen—a minute insect, for example—is stuck with a bit of glue onto a metal specimen stage about the size of a thumb tack, and the whole thing is put into an evacuated chamber. A sheet of gold alloy in the chamber is heated to a cherry red temperature. Atoms of gold boil off the sheet, float around through the jar, and come to rest on—among other things—the specimen. Within about an hour, the insect has received a uniform plating of gold a few atoms thick. If visible to the naked eye, it now looks like a beautifully detailed piece of jewelry. A large number of specimens can be plated simultaneously.

Since the circuits that control the picture produced by the scanning microscope are electronic in nature, they are extremely flexible. An operator can look over an entire specimen, find interesting parts, zoom in, change his angle of view, move from one part to another with ease. Thus a single operator can examine huge numbers of samples quickly and efficiently. All current instruments are designed with built-in cameras as well, so that when an operator sees a picture he wants, he simply pushes a button and the scene is recorded automatically.

With its combination of advantages, the scanning electron microscope is beginning to play an important role in many fields of scientific investigation. It will not replace either of the two standard microscopes; each has its own functions, which it performs superbly. In magnification or resolving power, the scanning instrument lies somewhere between the optical and transmission electron microscopes. It can resolve objects down to about 200 angstroms, as compared with 2000 for the optical microscope and two angstroms for the conventional electron instrument.

But with its unique surface-seeing capability, the new instrument will supplement the other microscopes in many important ways.

In the few years since it became available, the scanning electron microscope has produced much new knowledge, some surprises, and thousands of striking and frequently lovely pictures. Some of them appear on the following pages.

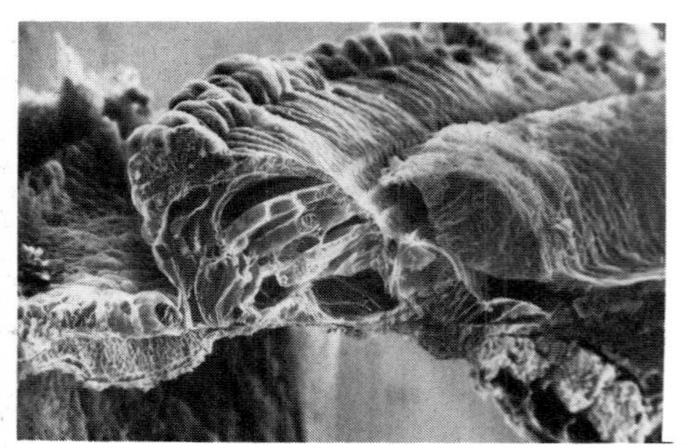

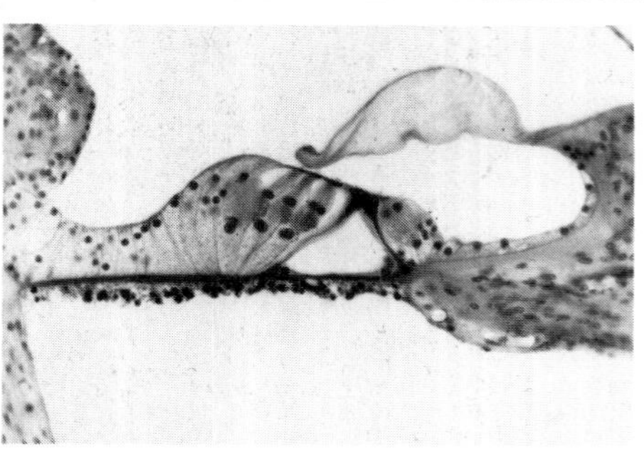

This picture pair demonstrates in a striking manner the different pictures obtained with a scanning electron microscope (top) and transmission instrument viewing the same subject. The scanning microscope, which reveals and photographs surfaces, shows a three-dimensional view of the small structure in the inner ear. By looking at such a picture, physiologists can see the overall structure of the organ and understand its physical function more fully. The transmission photo, on the other hand, is made from a thin cross section of the same organ. It does not tell nearly as much about the gross structure, but it reveals many internal details invisible in the other photo. Thus the two instruments, by revealing quite different aspects of the same subject, complement each other perfectly and lead to a far better understanding in biology, physiology, metallurgy, and many other fields.

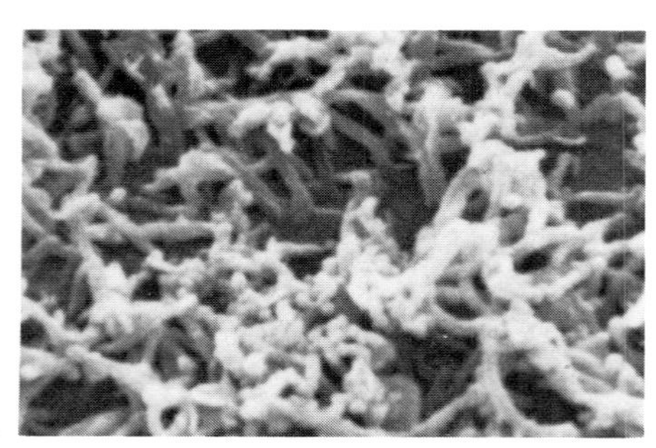

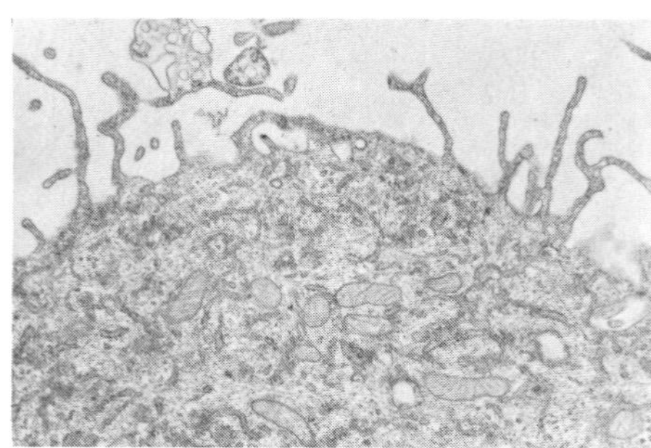

The transmission and scanning electron microscopes complement each other perfectly, each producing useful information. Both of these photos show a liver tissue culture cell. The transmission photo shows a highly detailed cross-section view, giving much detail on the cell's internal construction. But little can be seen of its outer surface. The scanning photo (top), on the other hand, shows the surface in vivid detail. The small fiber-like protuberances (microvilli), poorly seen in the transmission photo, turn out to cover the surface in the scanning microscope picture.

175 X (upper left), 136X (lower left)—Dr. Göran Bredberg, Akademiska Sjukhuset, Uppsala, Sweden

13400 X (upper right); 4000 X (lower right)—Dr. Patrick H. Cleveland, University of Minnesota

Magnifications follow each caption, with an abbreviated photograph credit. A more complete list of credits is included at the end of the book.

Although its operating principles are simple, the scanning electron microscope itself is a complex, massive instrument costing about $100,000. The specimen chamber is at the bottom of the column at left. A specimen is inserted through a small door, and pumps in the lower part of the instrument then create a vacuum in the column. This is necessary because the electron beam used to scan the specimen and create the image cannot pass through air; the electrons collide with air molecules and are deflected. Within the specimen chamber is a stage that can be twisted, turned, and tilted so that the specimen can be studied from any angle. The electronic circuitry for generating the electron beam, scanning it across the target, and creating the final picture is in the main unit. On the front panel are the operating controls. The panel also contains two identical viewing screens. The one at left is used for direct viewing as the operator manipulates the controls until he gets the precise picture he wants. Then he simply trips the shutter of the camera attached in front of the right-hand screen to record the scene.

CHAPTER 1

small creatures

Butterwort has a central stem with a blue flower on top and a half-dozen or so leaves two or three inches long which lie flat on the ground at the base of the stem. The scanning electron microscope reveals that the sticky-feeling leaves are actually covered with bulbous structures on tiny stalks. Each globular mass is a droplet of quick drying cement, about a tenth of a millimeter in diameter. Other smaller globule-like bits are scattered across the surface of the leaf.
425 X—Professor J. Heslop-Harrison, Dr. Y. Heslop-Harrison, Royal Botanic Gardens, Kew, England

Carnivorous Plant

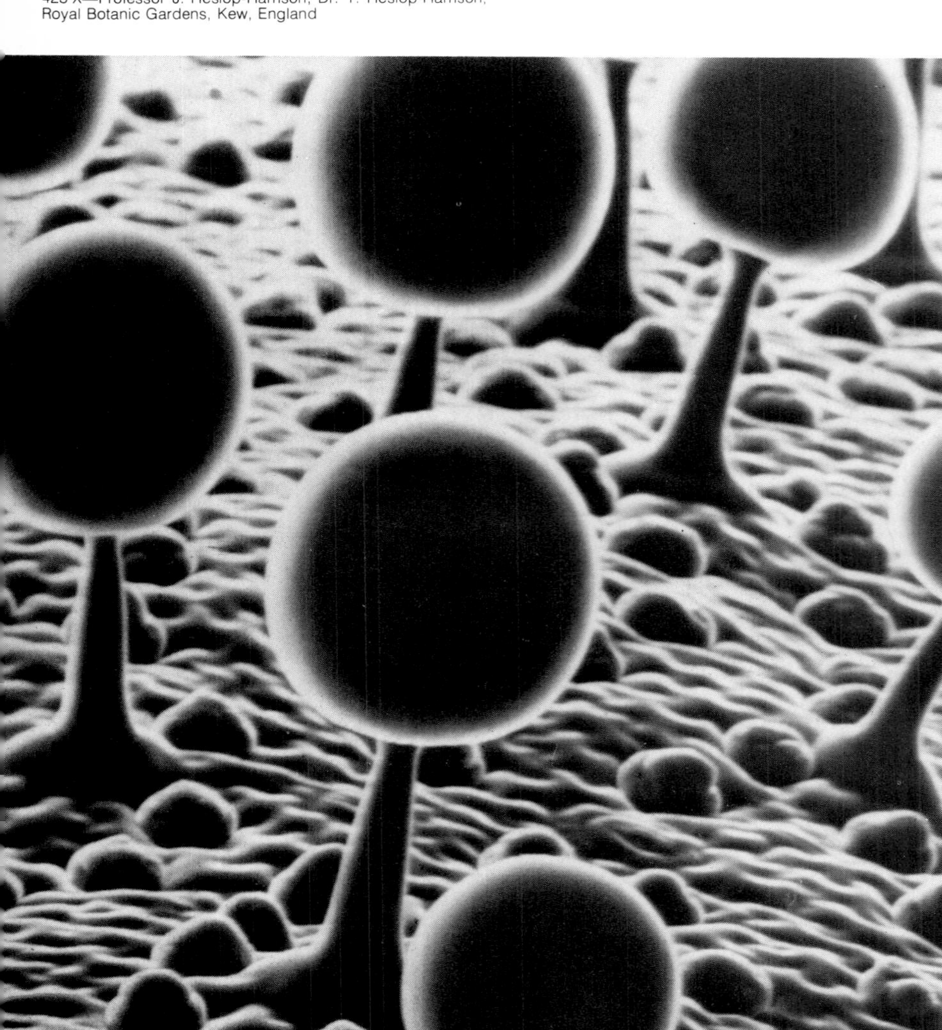

A plant captures its dinner. The butterwort plant (*Pinguicula grandiflora*) lives primarily in swamps and bogs, where the nitrogen supply is limited. So over the eons, it has gradually devised a peculiarly ferocious way of satisfying its need for this element and certain salts: it eats insects.

When an insect—an ant is shown here—touches ► the leaf, the globules stick to him. As he moves, the mucilage inside the globules stretches out into thin strands, then immediately hardens into restraining cables. The more the victim thrashes, the more globules he touches and the more strands tie him down. The ant and the plant seen here were living when this series of micrographs was taken; this was the first time that living, uncoated biological material had been studied with the aid of the scanning electron microscope. The ant was actually in its death struggles, and the plant glands were seen in the act of discharging their contents.
85 X—Professor J. Heslop-Harrison, Dr. Y. Heslop-Harrison, Royal Botanic Gardens, Kew, England

Here the ropes of hardened mucilage hold the ant ► fast. Some collapsed globules are in the foreground; others that have not been touched are as round and full as ever.
Once the capture is complete, the leaf forms a depression at the point where the insect's body is held prisoner. Then the small digesting glands which dot the surface of the leaf begin pouring out enzymes such as proteases, amylases, nucleases, and lipases, which are also found in the human digestive tract. These digest the insect rapidly, and released nutrients are absorbed into the leaf.
200 X—Professor J. Heslop-Harrison, Dr. Y. Heslop-Harrison, Royal Botanic Gardens, Kew, England

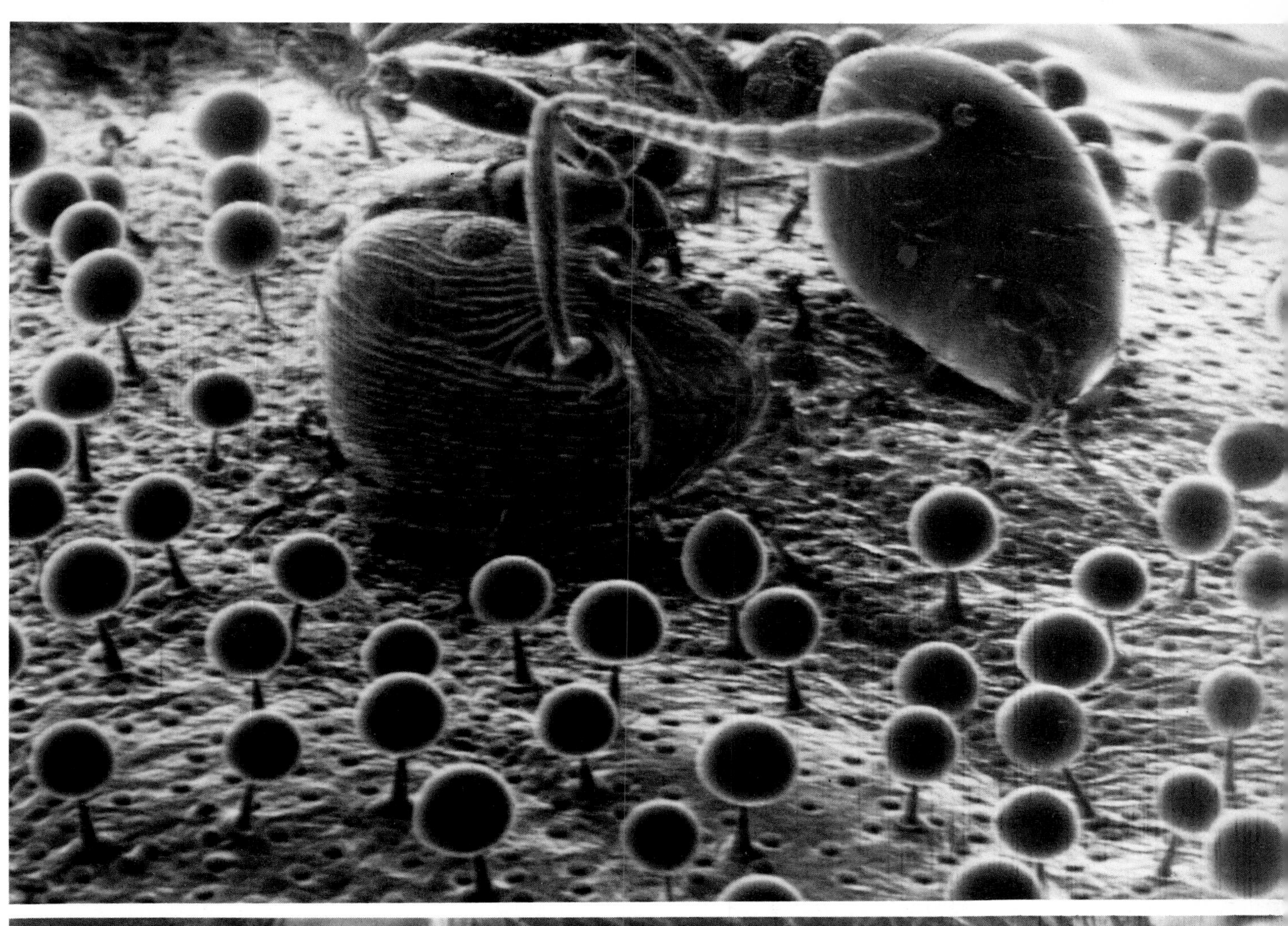

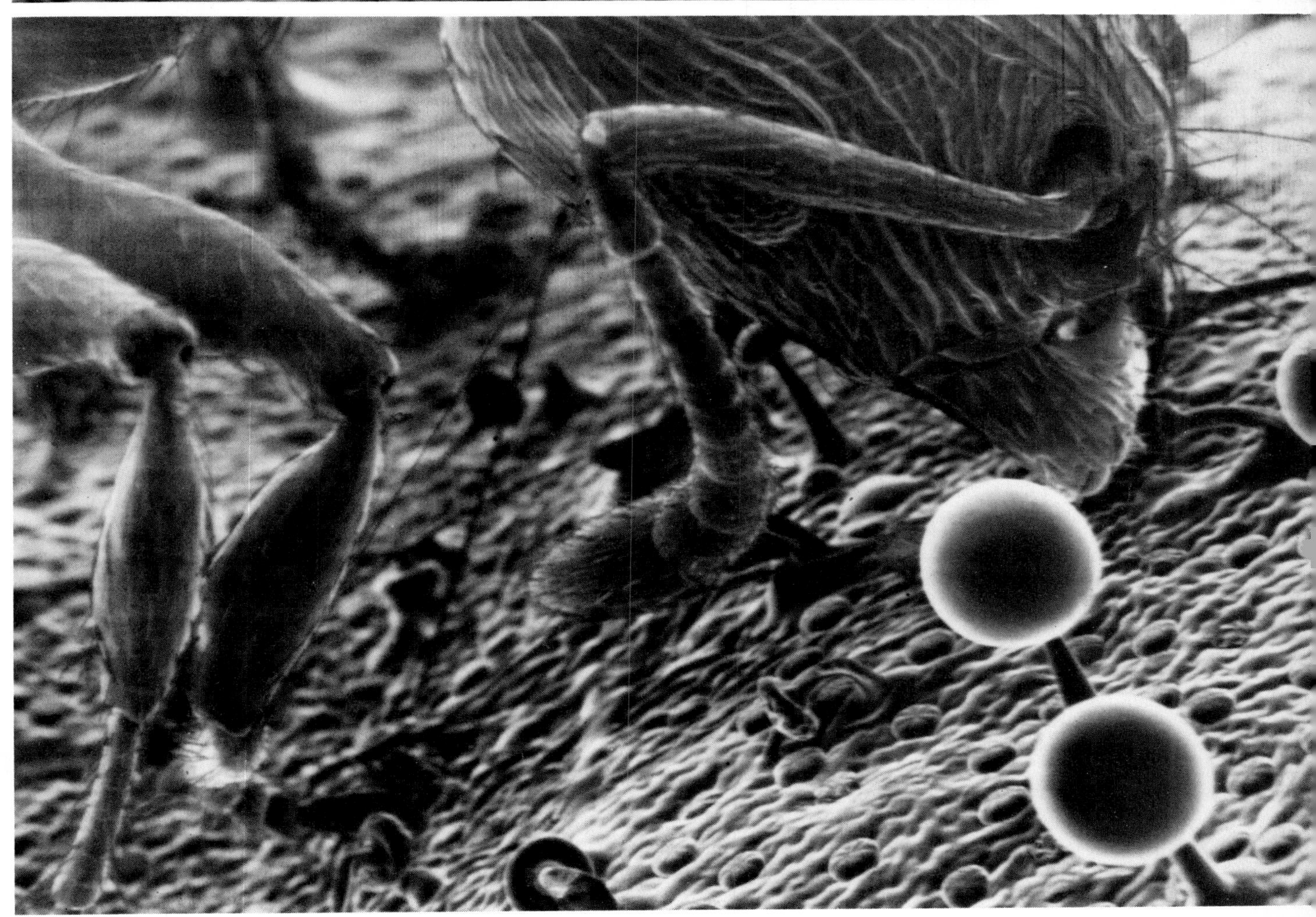

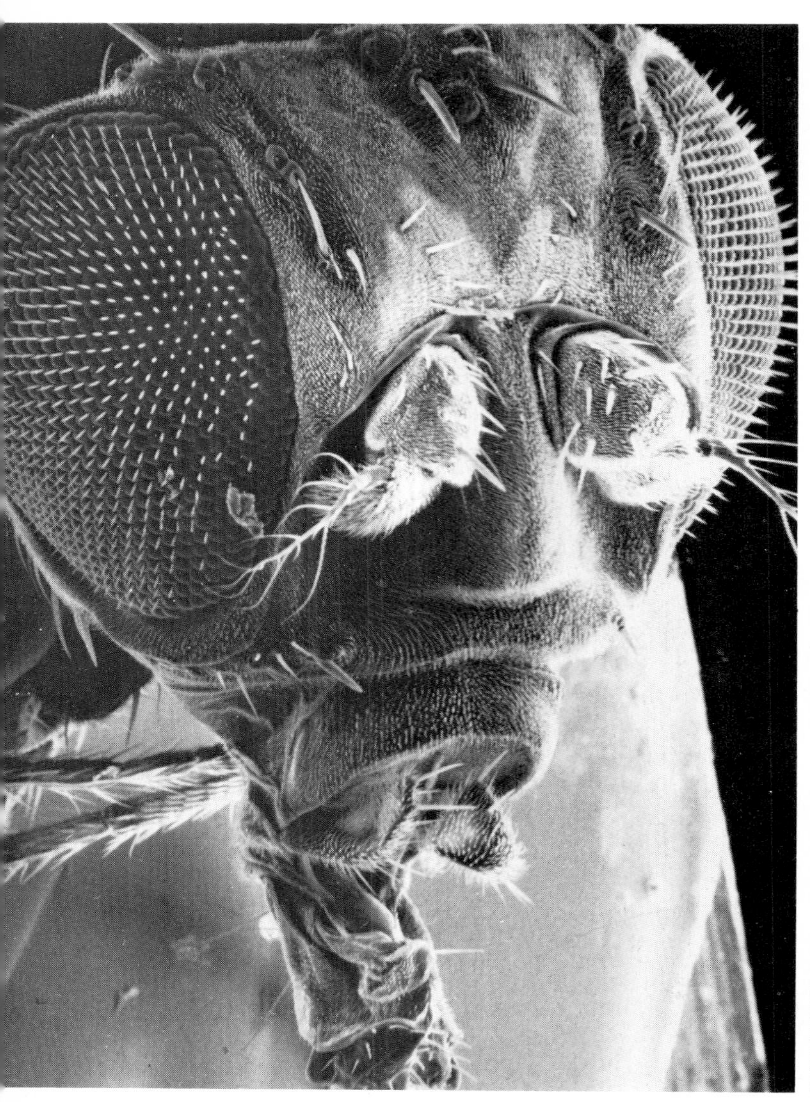

The fruit fly (*Drosophila melanogaster*), like most of the more than 75,000 known species of flies, is a fearsome looking creature when viewed close up. The multi-faceted eyes and the proboscis—used for sipping juice from fruits—are particularly prominent.
135 X—Dr. L. M. Beidler, Florida State University

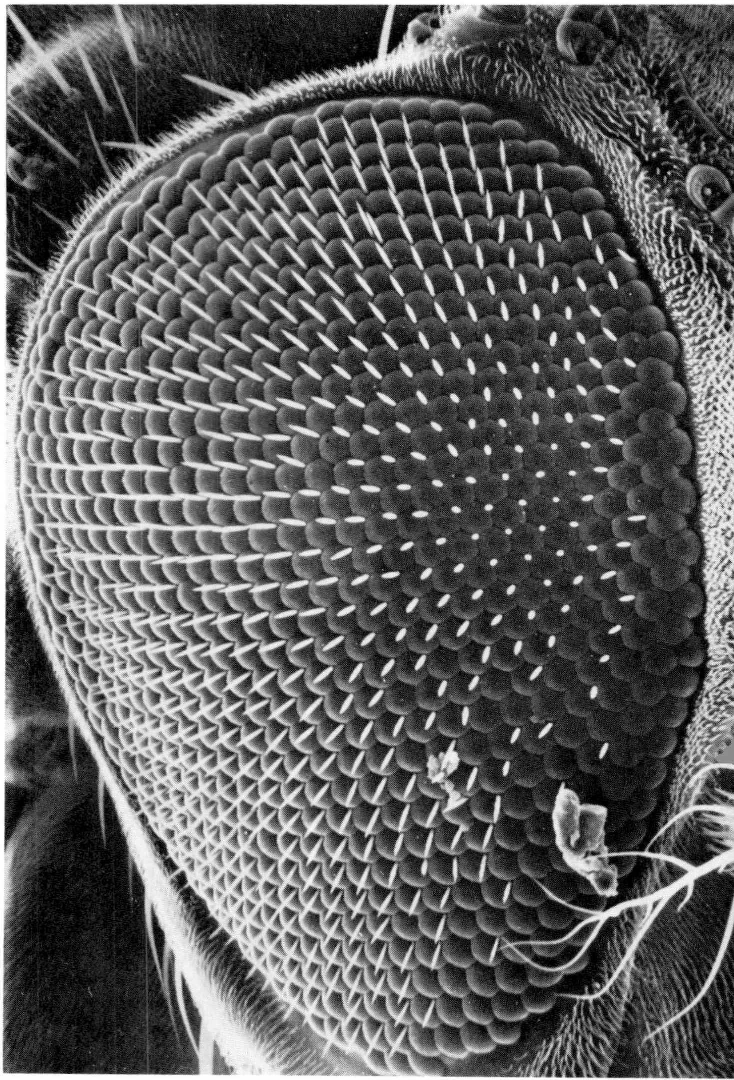

A closer look at the eye. Note the small bit of dust trapped on a hairlike appendage at bottom. Flies and many other insects have eyes made up of many simple lenses, or ommatidia.
260 X—Dr. L. M. Beidler, Florida State University

The facets, which contain corneal lenses, are beautifully grouped in regular rows. Tiny setiform hairs sometimes appear in the interspaces.
1450 X—Dr. L. M. Beidler, Florida State University

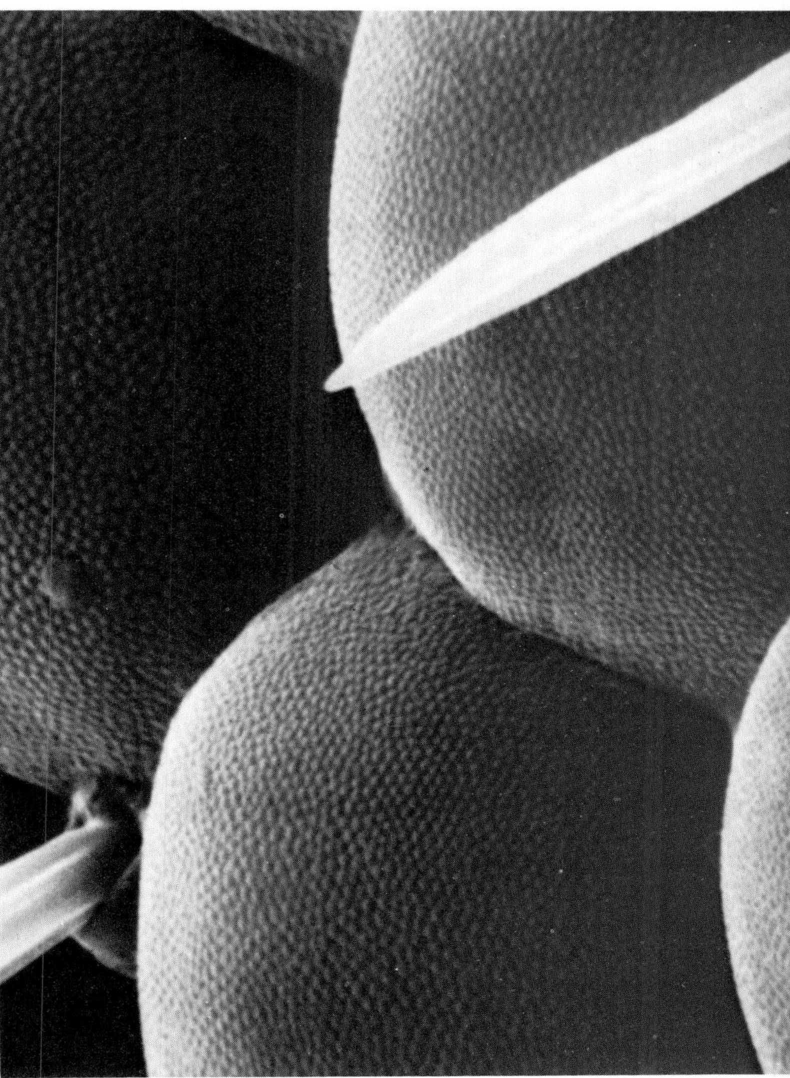

At great magnification, each eye facet appears to have a substructure consisting of small granular lumps distributed over its surface. The minute hairs are also seen to be fluted rather than cylindrical.
7800 X—Dr. L. M. Beidler, Florida State University

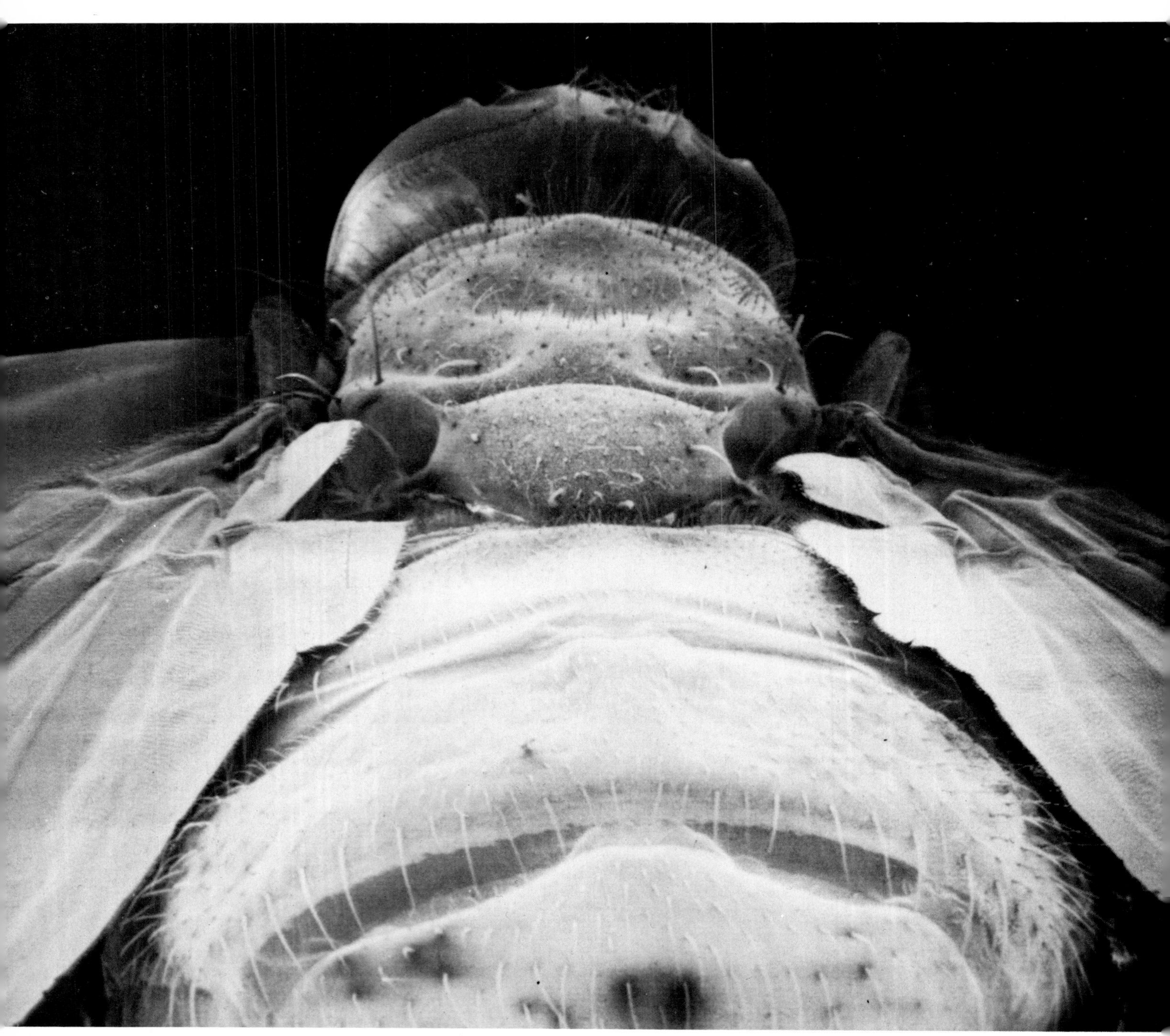

The scanning electron microscope shows the delicate, complex structure of the humble stable fly in this view looking up the back toward the head. The graceful and functional folds of the wing give strength and rigidity to an extremely light structure. The body is covered with fine hairs.
80 X—Johnson Wax Photo

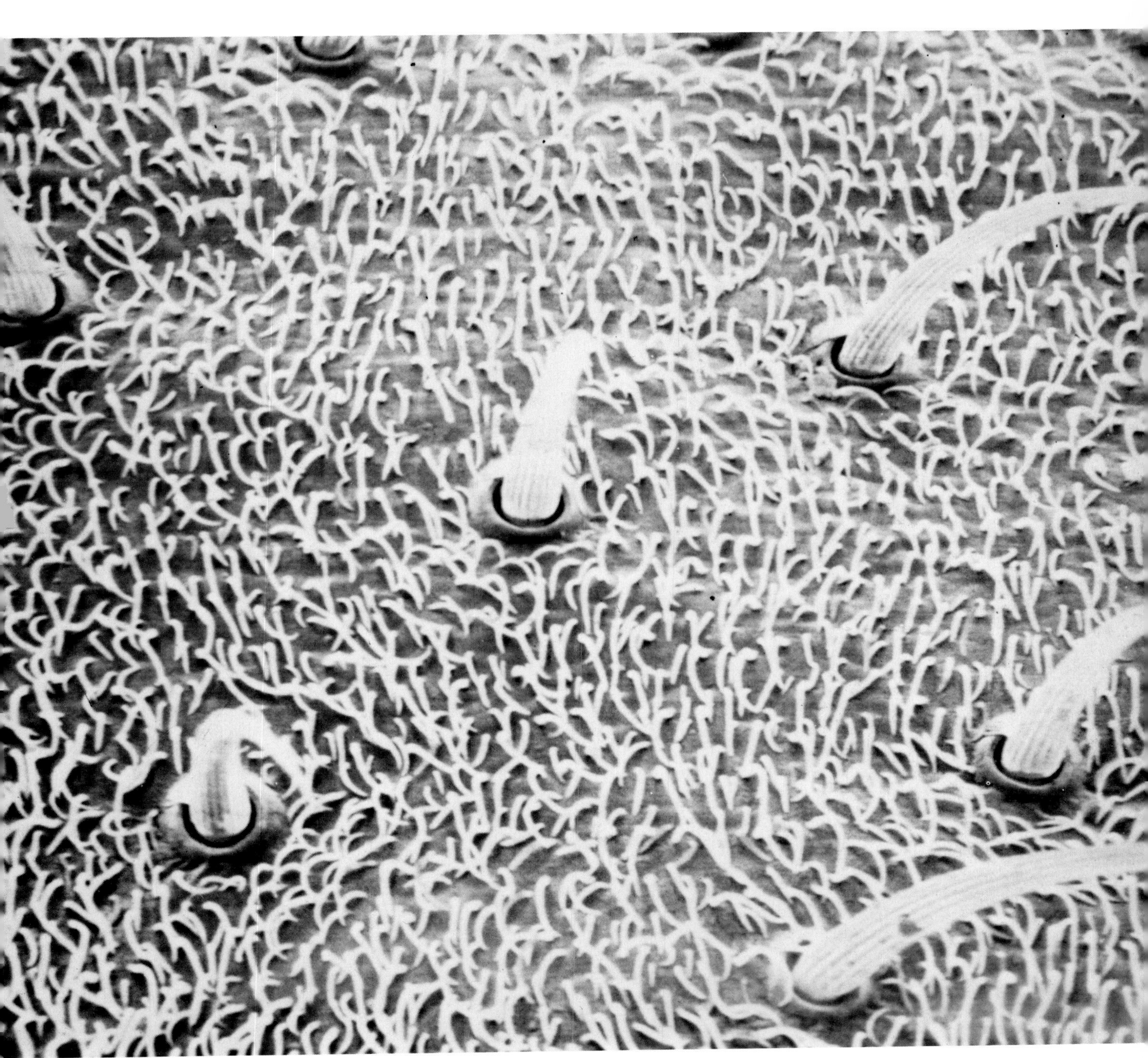

A closer look reveals that the fine hairs seen in the previous photograph are fluted and sprout from neat, doughnut-like structures. And they are surrounded by a field of yet smaller hairs.
1500 X—Johnson Wax Photo

The tsetse fly transmits sleeping sickness and is a dreaded pest in parts of Africa south of the Sahara. Yet this photograph reveals him to be a fascinating looking creature. The proboscis, large for a fly but indispensable for one which lives by piercing the skin and sucking the blood of warm-blooded creatures, is at bottom, pointing toward the viewer. The eyes are the two huge oval structures on the sides of the head; small individual facets are just visible. The hair on the lower part of the head resembles mutton-chop whiskers.

70 X—Eastman Kodak Company, Industrial Laboratory

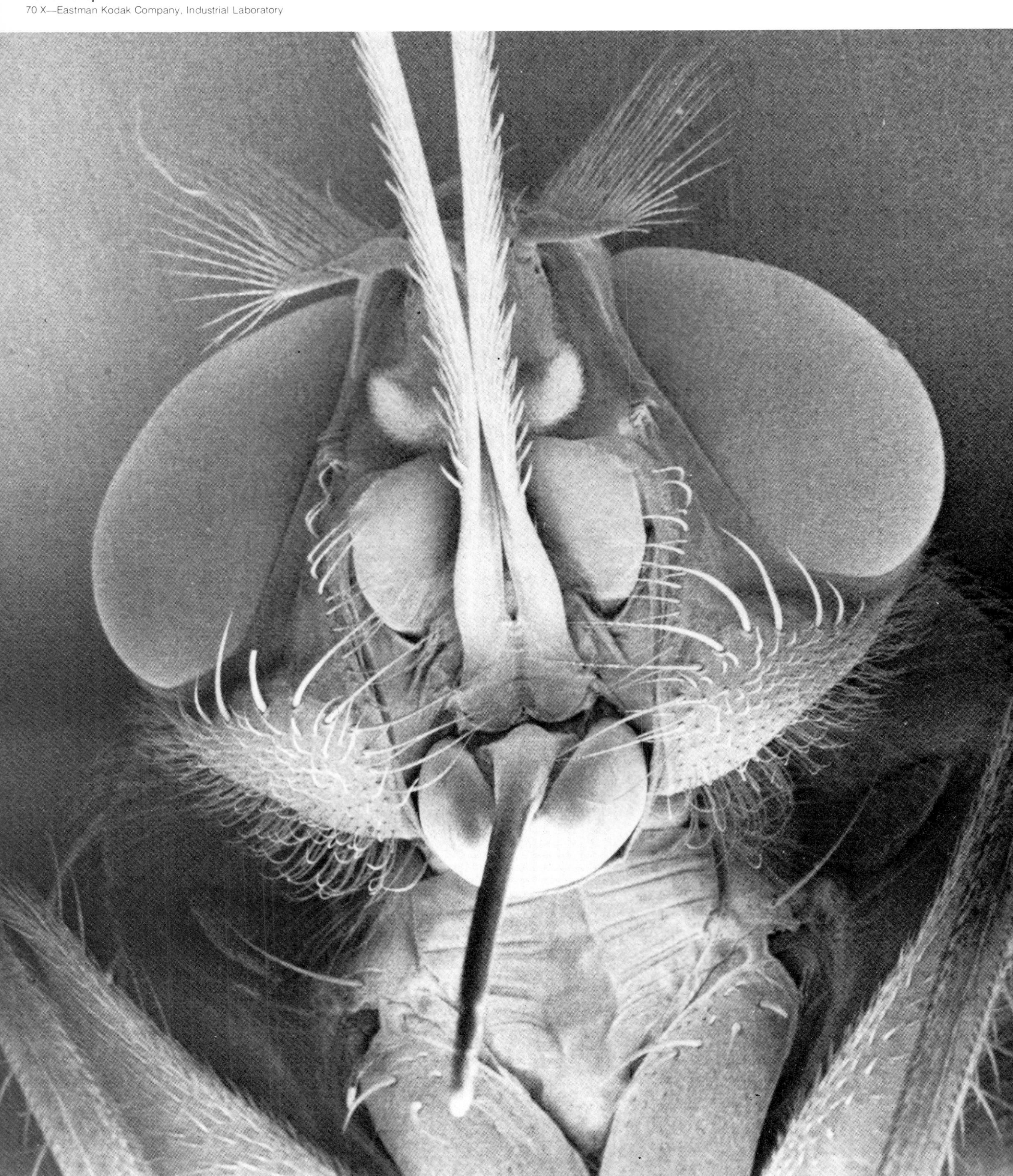

The previous picture showed the tsetse fly's proboscis as a single needlelike probe. Greater magnification reveals marvelous details of structure: the taper of the end, the bifurcated character, the many small auxiliary barbs.
1500 X—Eastman Kodak Company, Industrial Laboratory

The tsetse fly's foot is a marvel of design. Note the long, tapered hairs with their spiraling flutes, the joints, and the fringelike pattern on the footpads.
190 X—Eastman Kodak Company, Industrial Laboratory

A closer look at the foot shows that the footpad fringe is more than decorative detail. Each tube ends in a small suction pad—which accounts for the fly's ability to walk in any position on almost any surface.
1900 X—Eastman Kodak Company, Industrial Laboratory

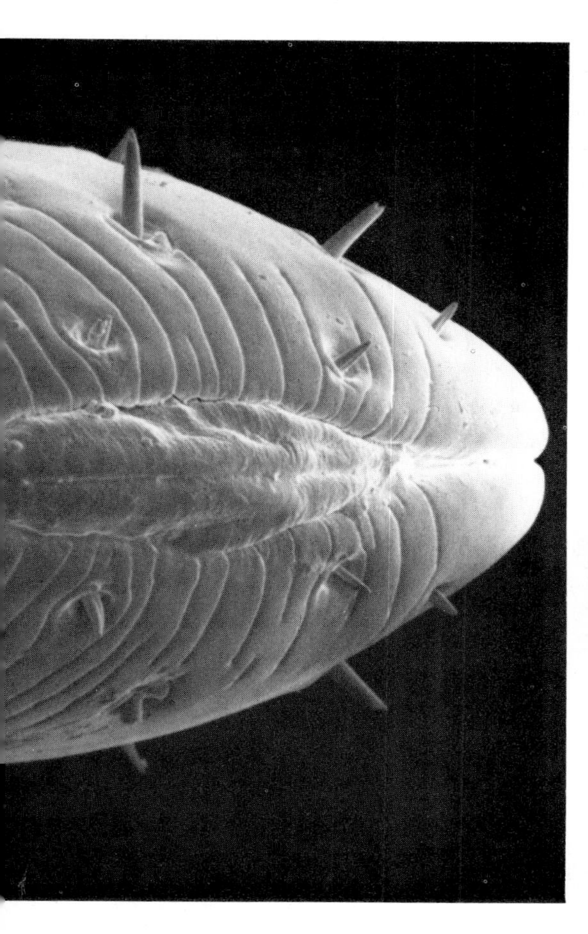

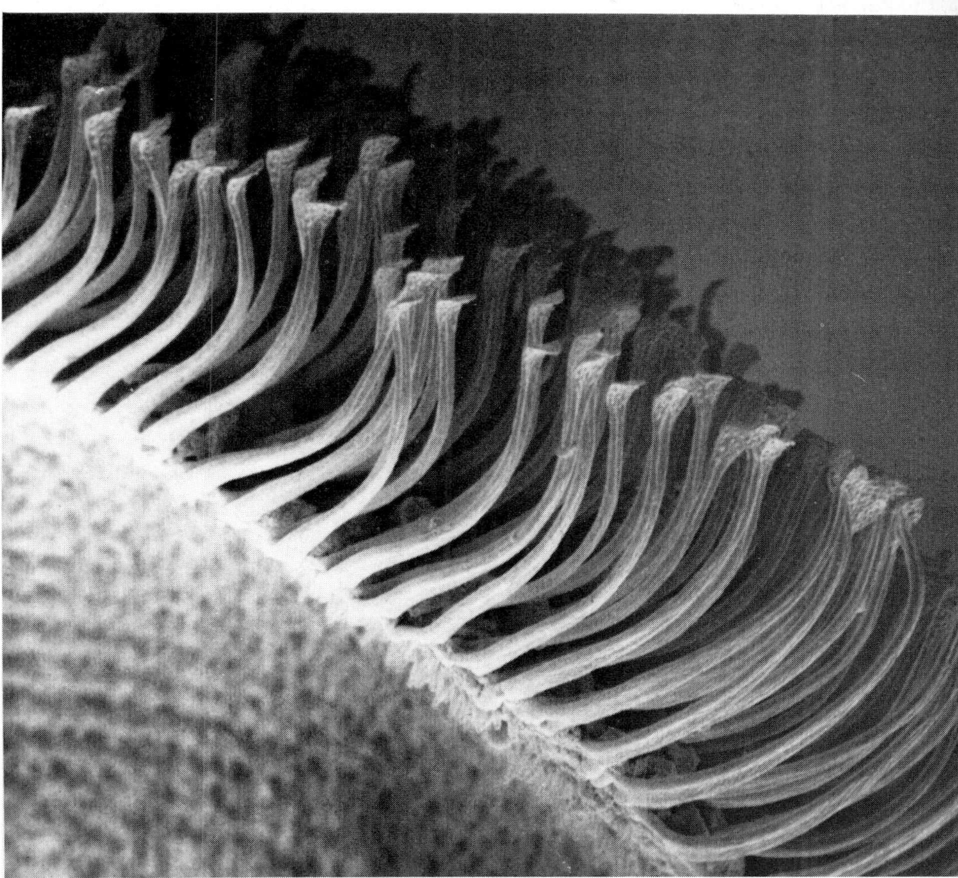

Ants

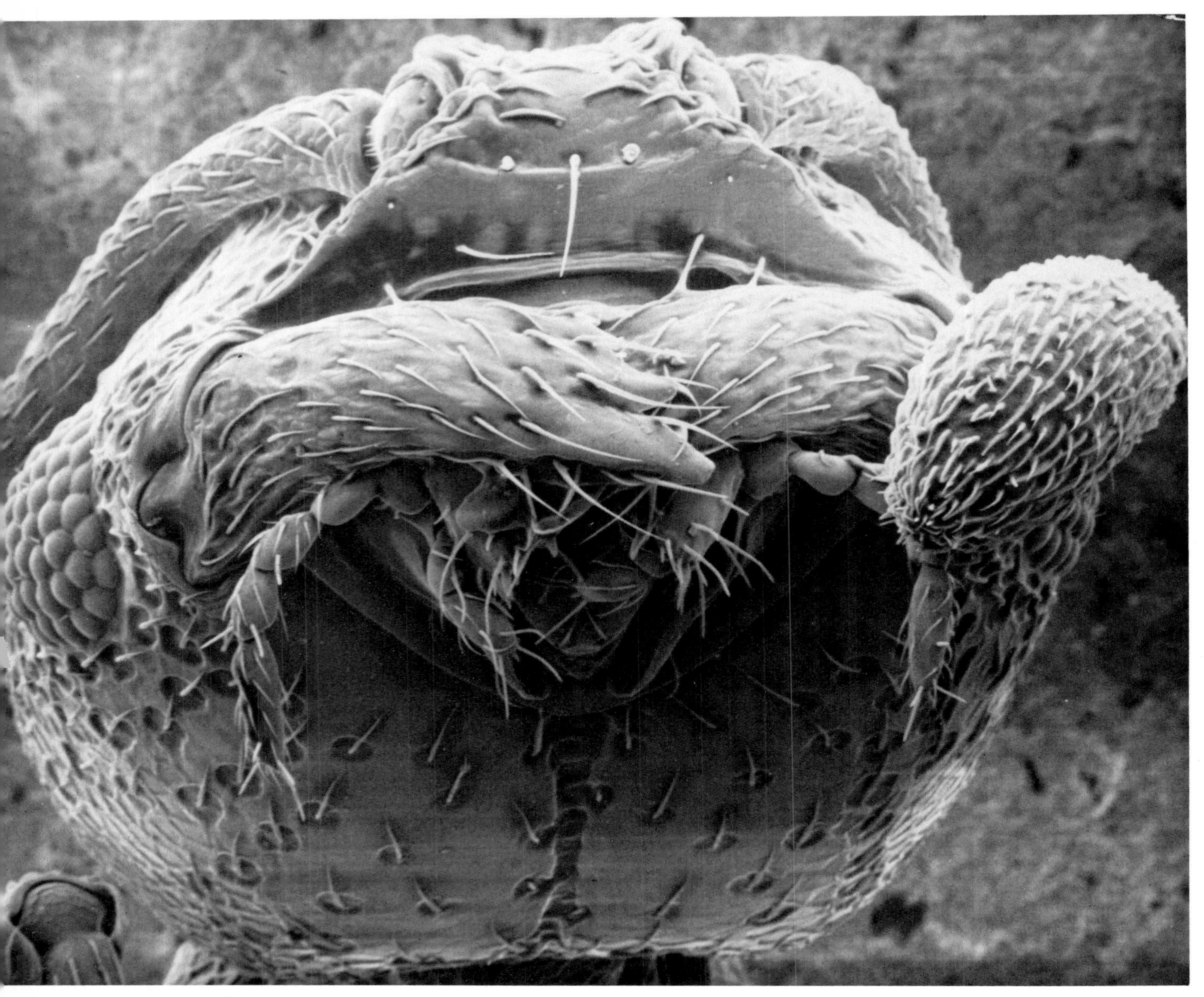

Crawling across a picnic lunch, one ant looks pretty much like another. But few domains of the animal kingdom show as much variety as that of these highly social insects.

Cardiocondyla wroughtoni foral. **Magnificent detail in this head-on view. That's the tip of an antenna coming around from the right. The mouth parts are at the bottom front of the head, and one eye—looking like a raspberry—can be seen on the side of the head.**
500 X—Eastman Kodak Company, Industrial Laboratory

Profile of the same ant. The location of antennae and eyes can now be easily seen. Note that each hair grows out of the center of a nearly circular depression. ►
390 X—Eastman Kodak Company, Industrial Laboratory

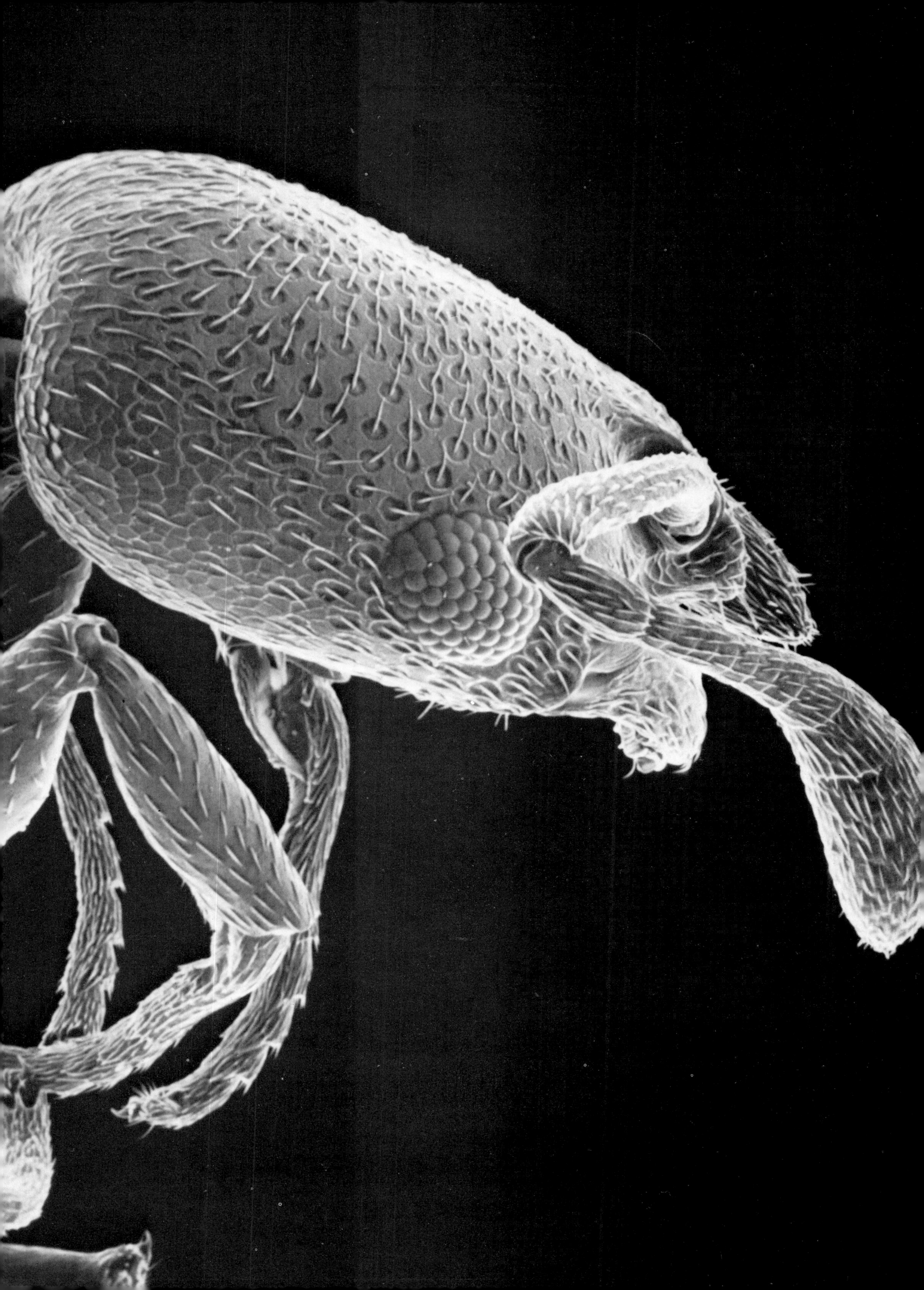

The antenna and eye structures are similar, but the head of this ant differs sharply from that of *Cardiocondyla wroughtoni foral.* There are no circular depressions and fewer hairs. Zoologists identify members of the ant world by their constricted waist and elbowed antennae; one can clearly see the form of the latter here.
60 X—Dr. L. M. Beidler, Florida State University

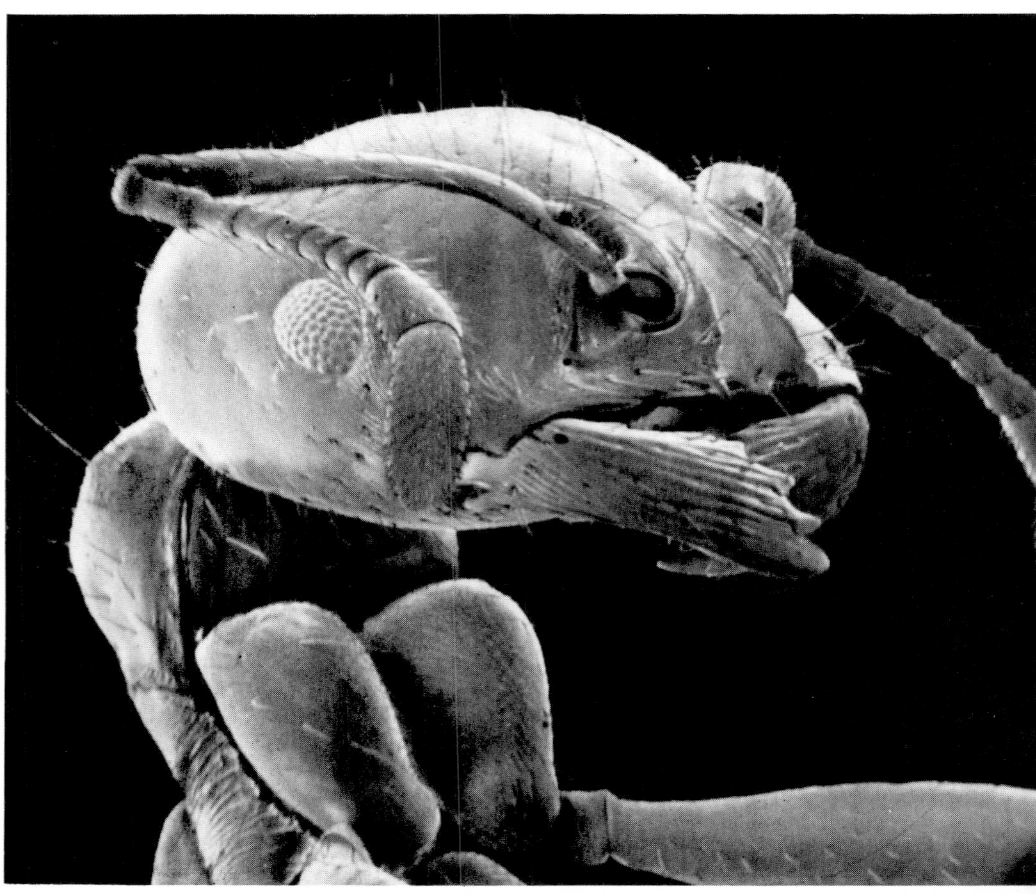

The mouth parts of most ants are made up of structures for biting and for taking up liquid food. Here the biting members with their serrated edges are seen in striking detail.
100 X—Dr. L. M. Beidler, Florida State University

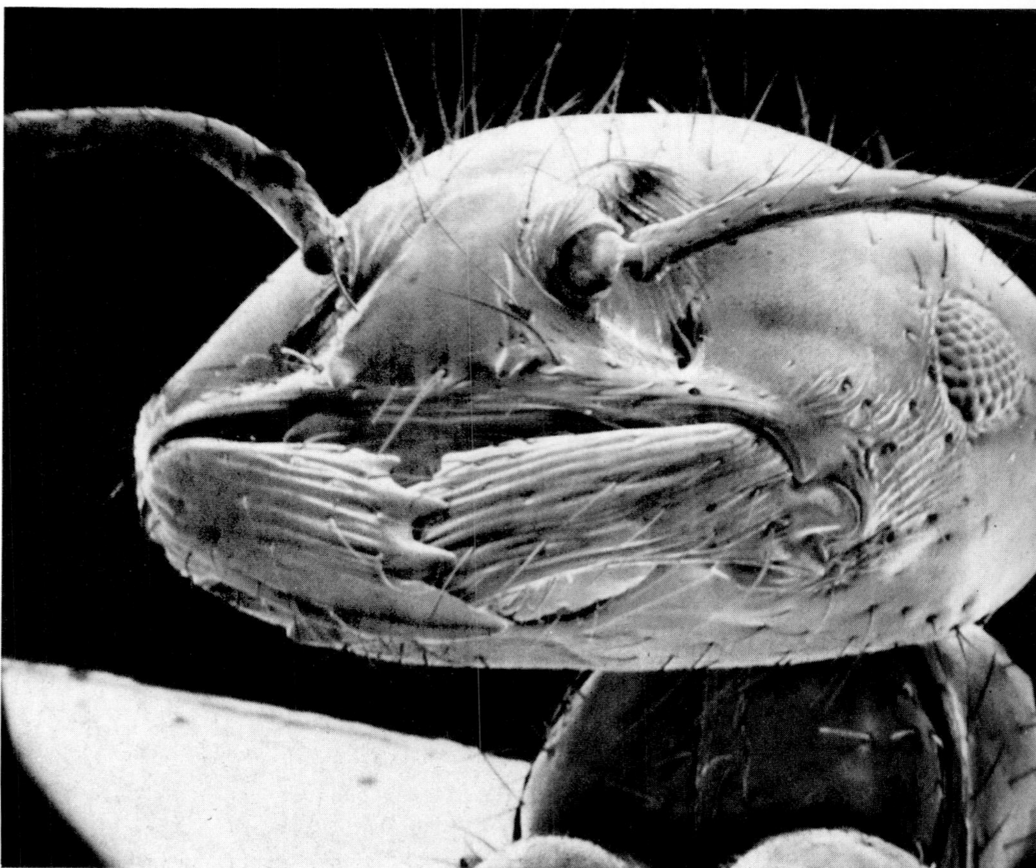

This may be the weirdest ant of all. *Cryptocerus Texanus Santachi* has a flat, wedge-shaped head with a highly textured surface. An eye can be seen at left.
85 X—Eastman Kodak Company, Industrial Laboratory

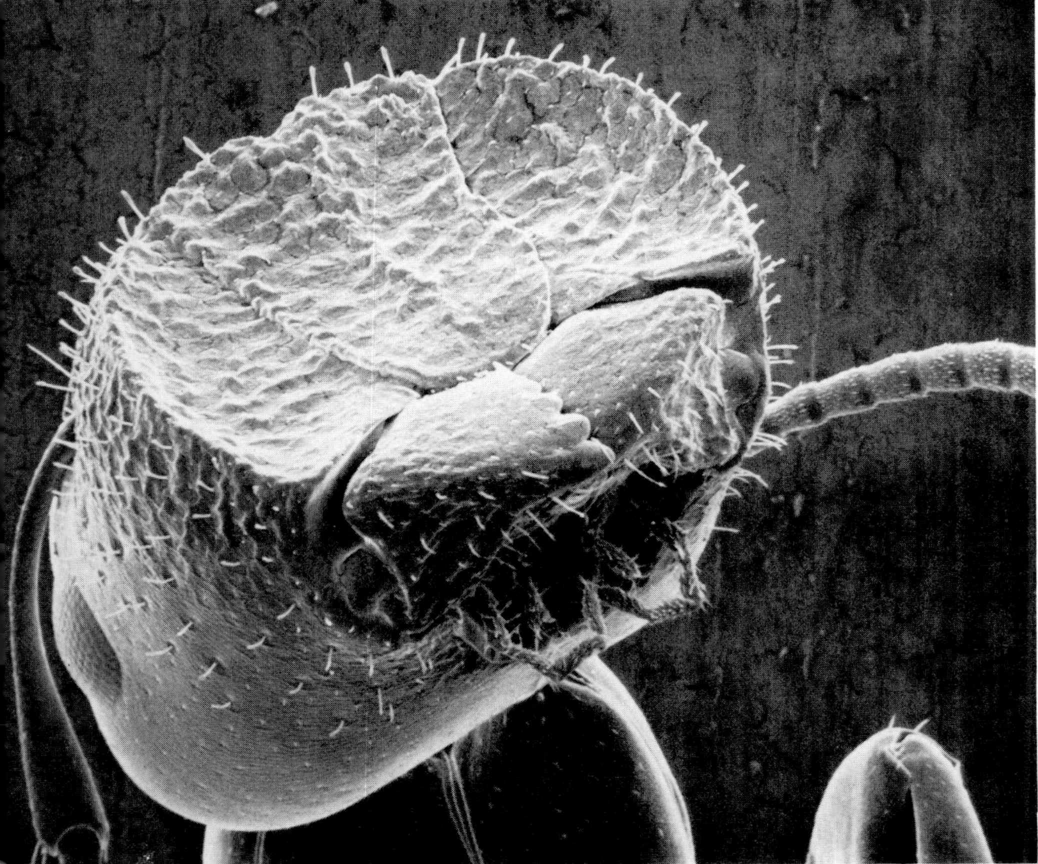

***C(Colobopsis) etialatus* looks as though he collided with something, but that's simply the way eons of evolutionary development have shaped him to survive. An antenna can be seen arcing over the eye at left.**
90 X—Eastman Kodak Company, Industrial Laboratory

Gnats

Leptoconops kerteszi Kieffer, no bigger than the head of a pin, is said to be the most ferocious creature for its size on earth. A team studying the gnat for the U.S. Army found that the tiny creature inflicts a more painful bite than his larger cousin, the mosquito, breeds by the billions in the world's arid regions, will fly up to ten miles, is attracted by carbon dioxide (and thus to anything that exhales), and is deterred by no known repellent.
75 X—Dr. Don M. Rees, University of Utah

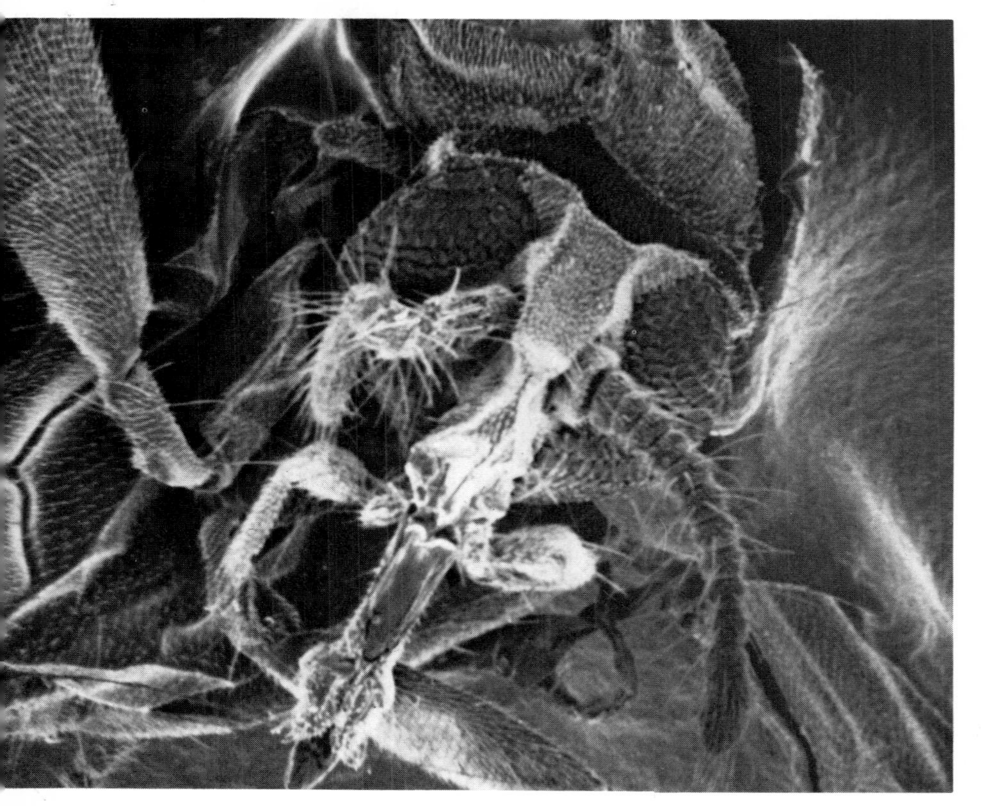

As with the mosquito and many other bloodsucking insects, it is the female gnat who performs the act man considers most anti-social. This view of the gnat's face shows the sharp proboscis with which she pierces human skin.
100 X—Dr. Don M. Rees, University of Utah

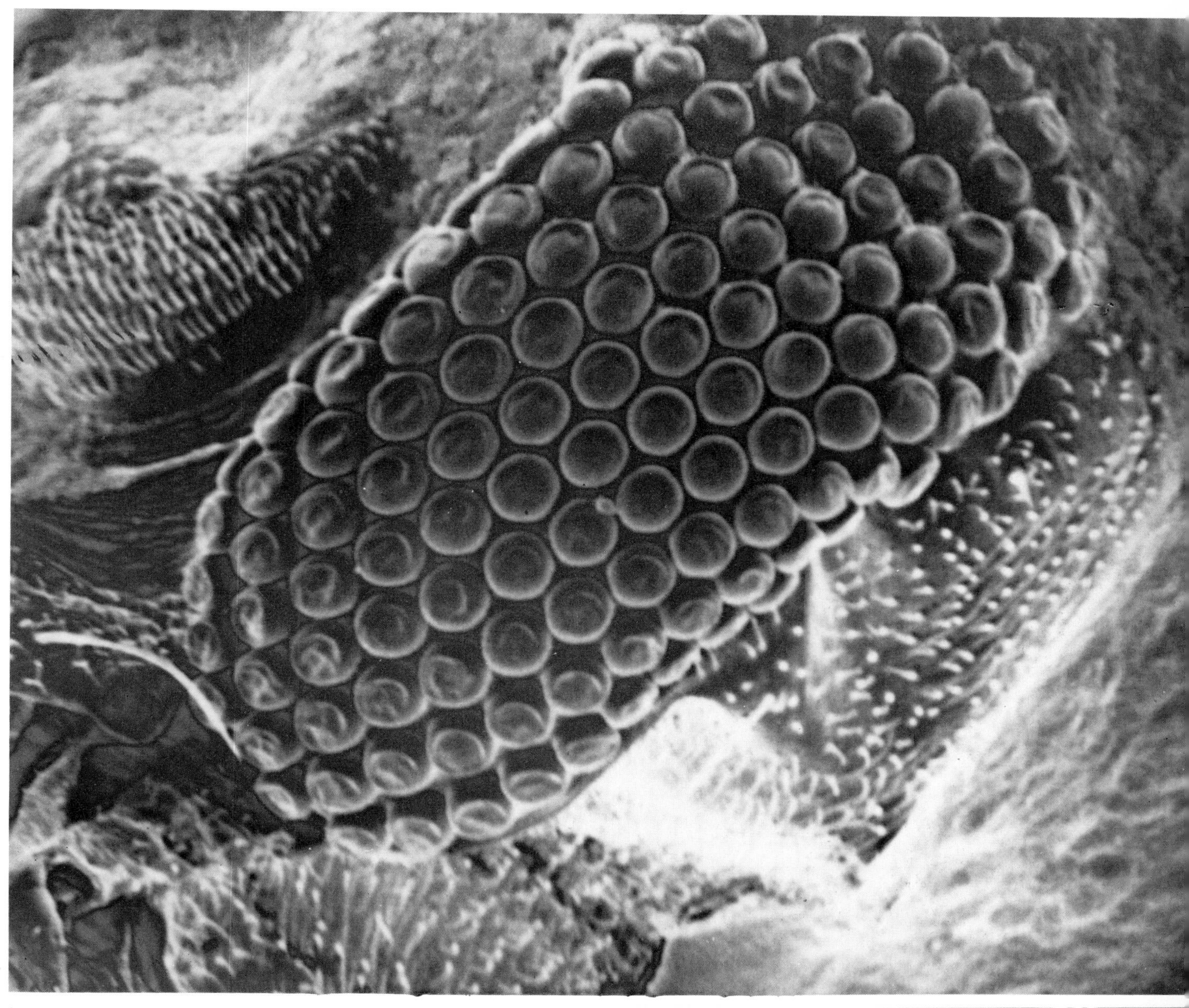

***Leptoconops kerteszi's* eye shows the usual multi-faceted form.**
400 X—Dr. Don M. Rees, University of Utah

Here is a greatly enlarged view of the needle-sharp proboscis tip that punches easily through human skin.
1100 X—Dr. Don M. Rees, University of Utah

Mosquitoes

Gnats and mosquitoes are both really members of the fly family, order *Diptera*. Therefore it is not surprising that they have many similarities. This head-on view of a mosquito has the same general appearance as a gnat's.

35 X—Eastman Kodak Company, Industrial Laboratory

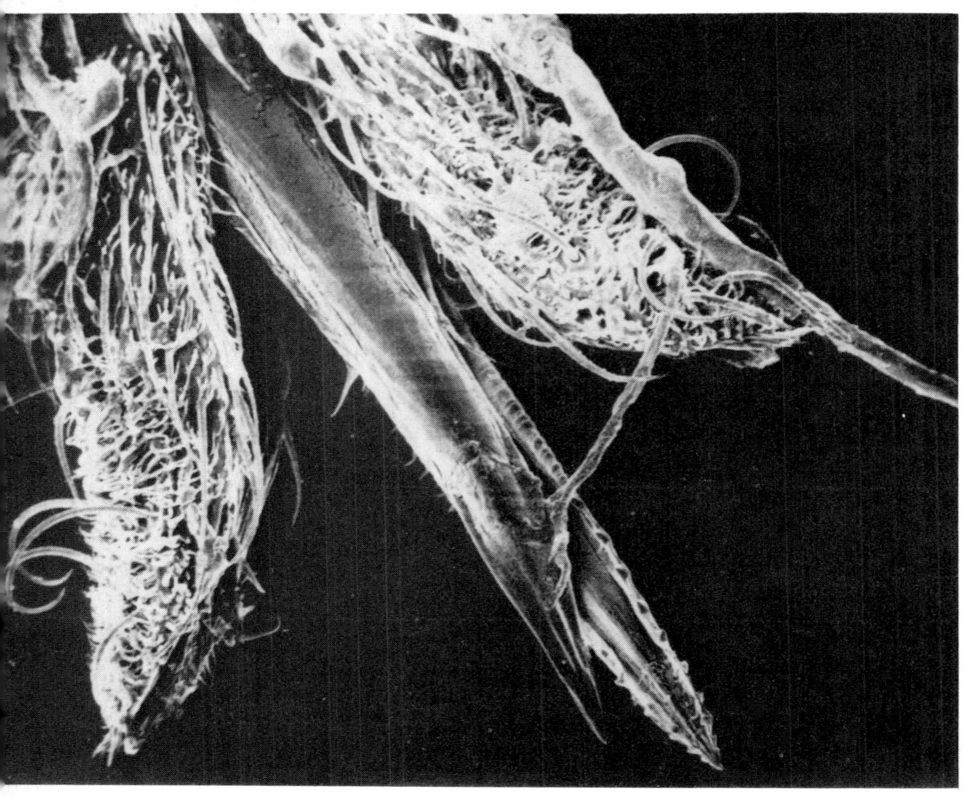

The end of the mosquito's proboscis is also similar to that of the gnat. The hypodermic-like tube pierces the skin and serves as a conduit for the insect's meal of blood. The complex structure around the central tube ejects a fluid that breaks down blood cells and prevents coagulation as the mosquito eats. Later its fine hairs may reabsorb much of the fluid. If they do, the victim suffers very few after-effects from the bite.

800 X—Eastman Kodak Company, Industrial Laboratory

The proboscis of another breed of mosquito shows the marvelously detailed structure even better. The passageway for blood can be clearly seen.
2300 X—Dr. L. M. Beidler, Florida State University

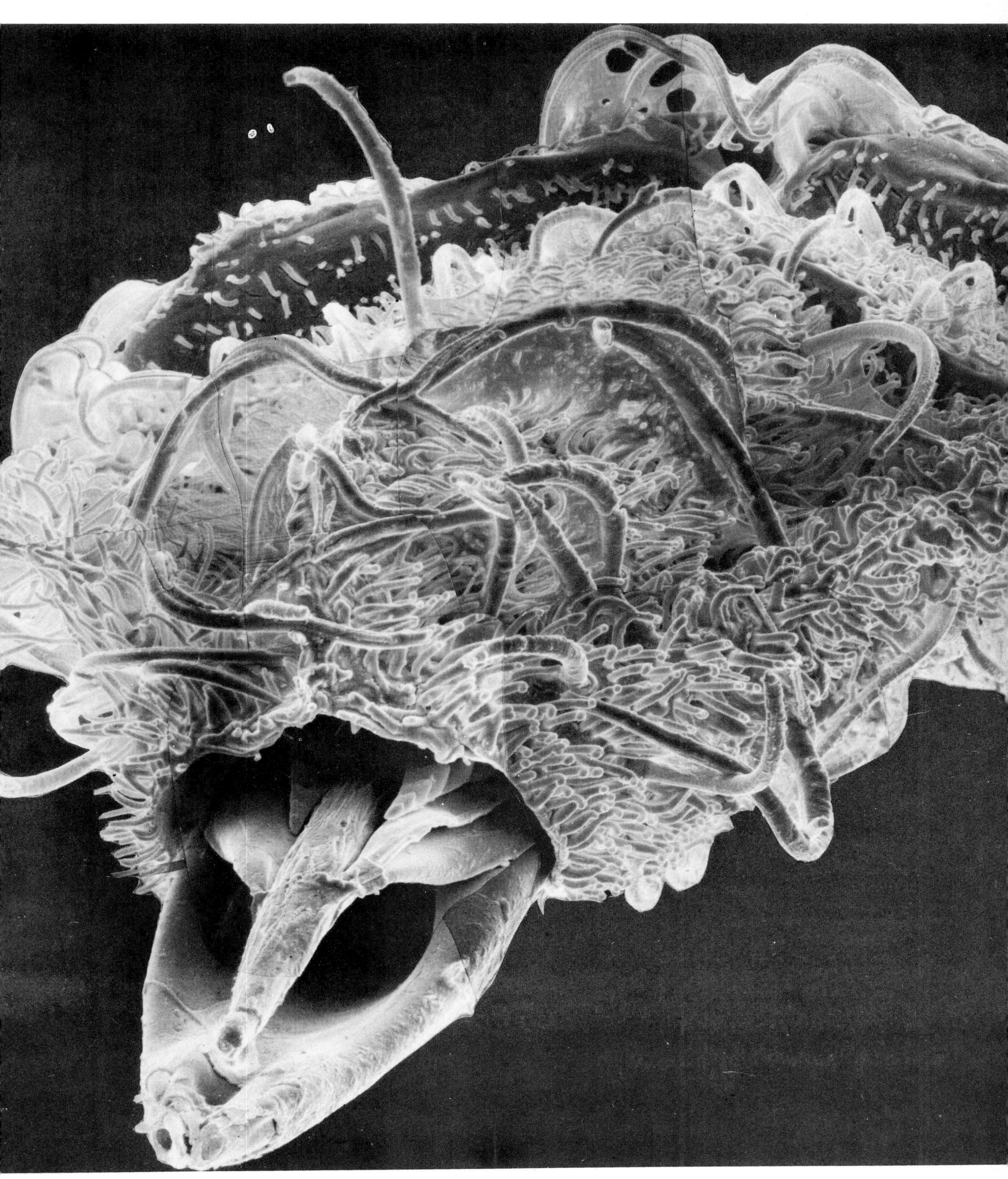

◀The bee belongs to an order called *Hymenoptera*, a completely different insect tribe from gnats and mosquitoes. He also uses his stinger for a completely different purpose. Thus it is not surprising that it in no way resembles the proboscis of a gnat or mosquito. This photo reveals that nature's handiwork is better than man's—at least in this instance—because unlike the eye of the needle through which it sticks, the bee's stinger is a masterpiece of precise construction.
180 X—Eastman Kodak Company, Industrial Laboratory

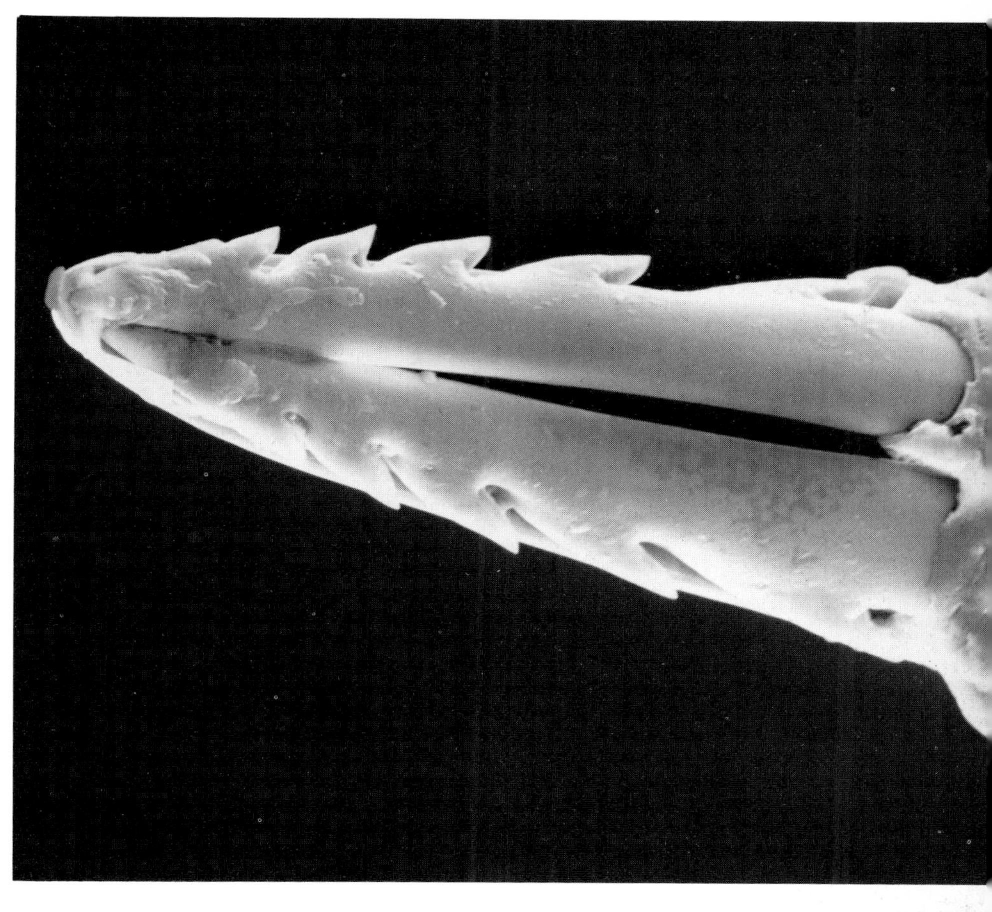

Once embedded, the stinger is difficult to extract. Here's why.
1100 X—Eastman Kodak Company, Industrial Laboratory

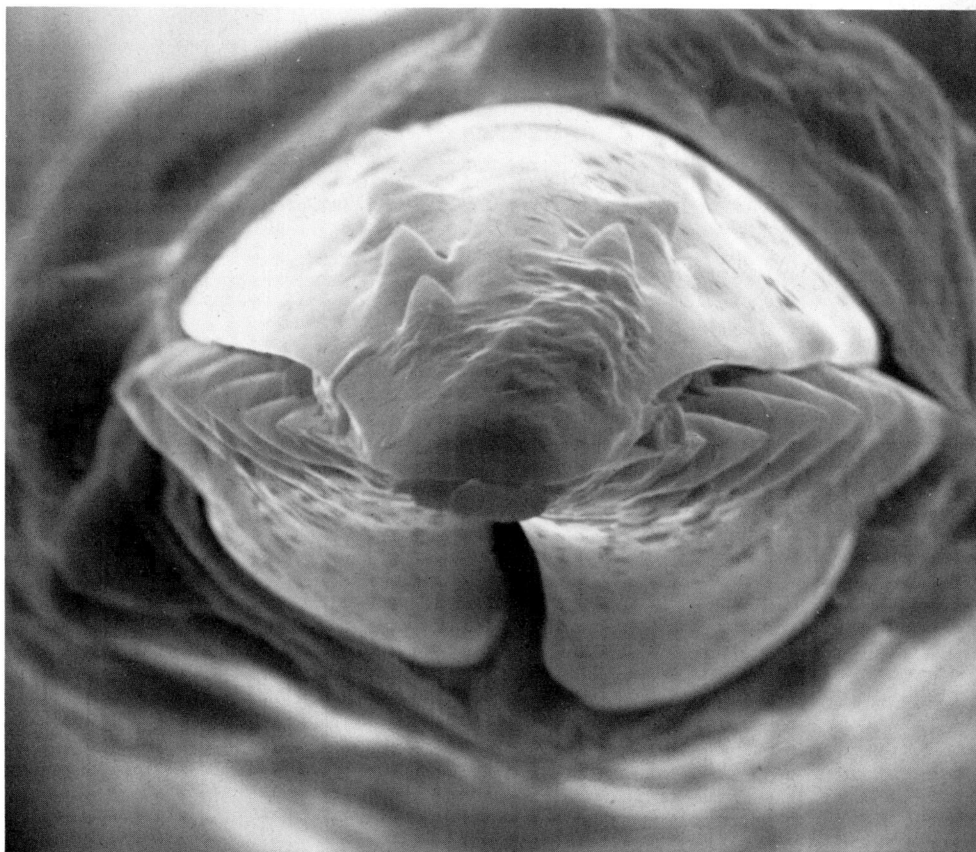

Skin's-eye view.
1850 X—Eastman Kodak Company, Industrial Laboratory

Butterfly Wings

Colias eurytheme is an attractive butterfly whose orange wings are bordered by a black band fringed with pink. The wings are made of tiny scales, each of which exhibits a delicate pattern. Further, the scanning electron microscope shows that orange scales differ not only in color, but also in structure from black ones, and the pink scales resemble neither.

This photograph shows how the orange scales overlap. Each scale is approximately a tenth of a millimeter in length and shows a distinct rib pattern.
1400 X—Annemarie C. Reimschuessel, John M. Kolyer, Allied Chemical Corporation

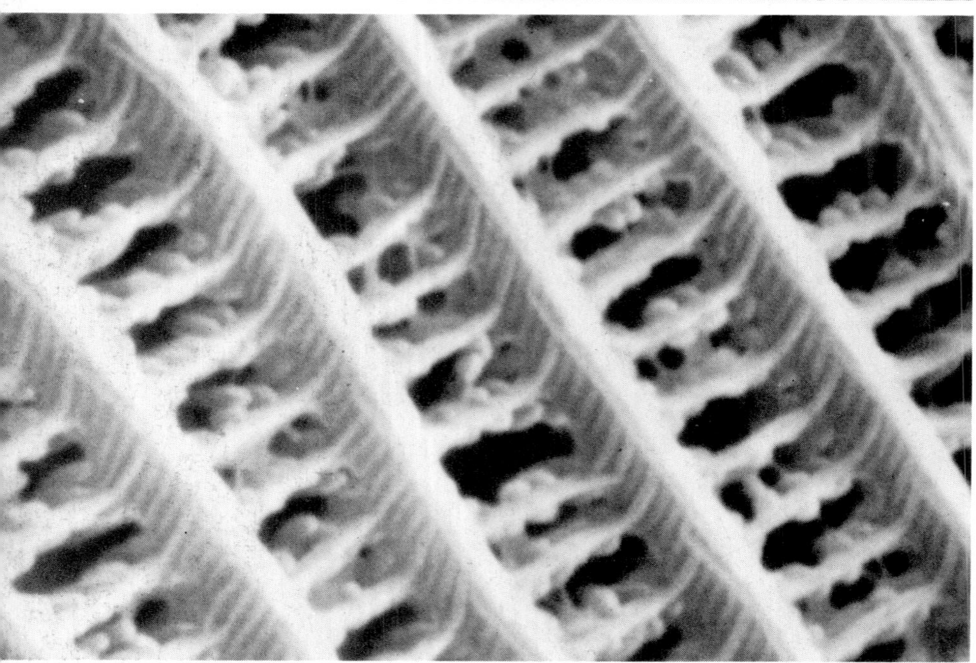

A closer look at the scale shows that the ribs apparent in the previous photographs are connected by a series of cross-struts. The combination forms a light, strong structure.
14,000 X—Annemarie C. Reimschuessel, John M. Kolyer, Allied Chemical Corporation

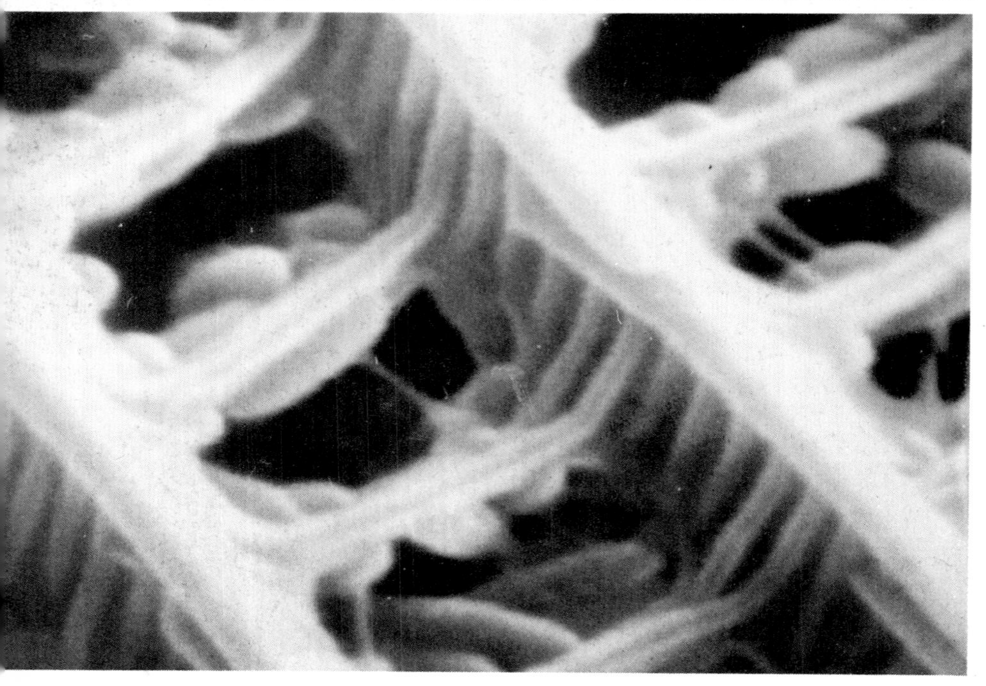

Still greater magnification reveals curious ellipsoidal structures attached to the cross-members. Their function is unknown, but some investigators suspect they may be color pods, associated with the scale's hue.
40,000 X—Annemarie C. Reimschuessel, John M. Kolyer, Allied Chemical Corporation

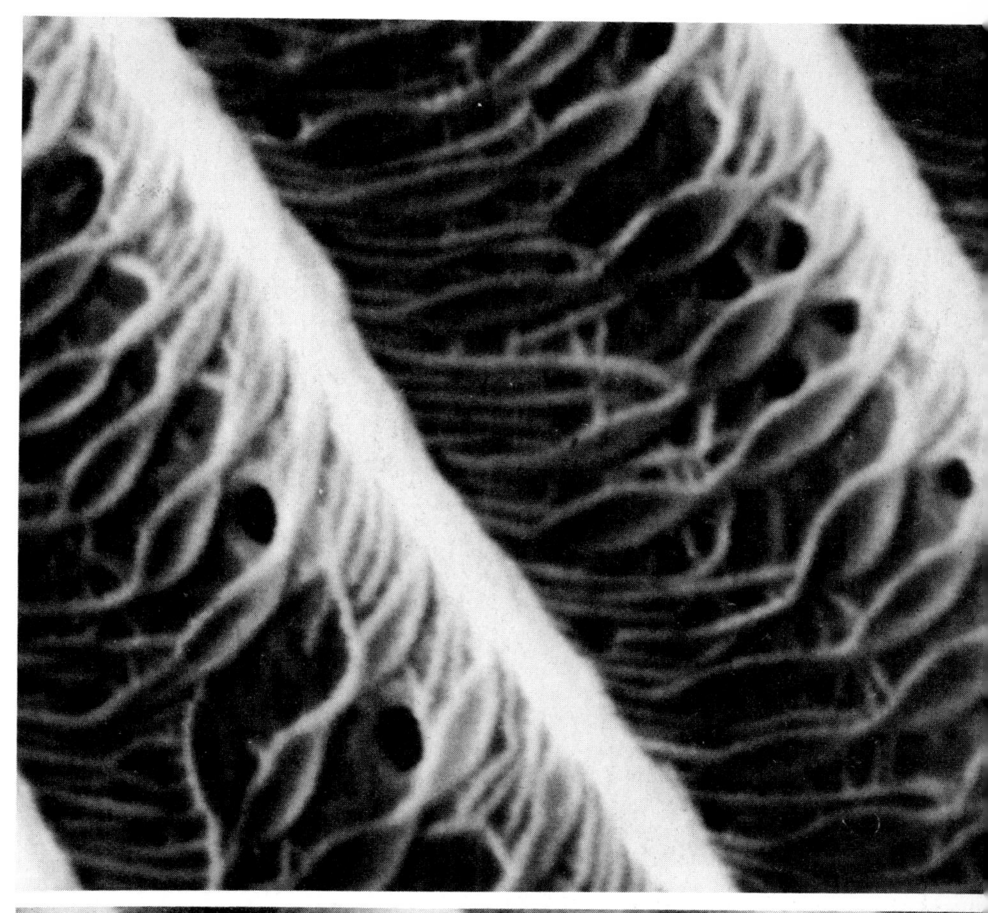

The structure of the upper surface of a black scale is strikingly different, with a series of curving, double-lobed struts connecting the ribs.
14,000 X—Annemarie C. Reimschuessel, John M. Kolyer, Allied Chemical Corporation

Yet another variation in the pink fringe scales. There must be a reason for the startling differences in structure between scales of different color, but no one knows what it is.
16,000 X—Annemarie C. Reimschuessel, John M. Kolyer, Allied Chemical Corporation

Mites

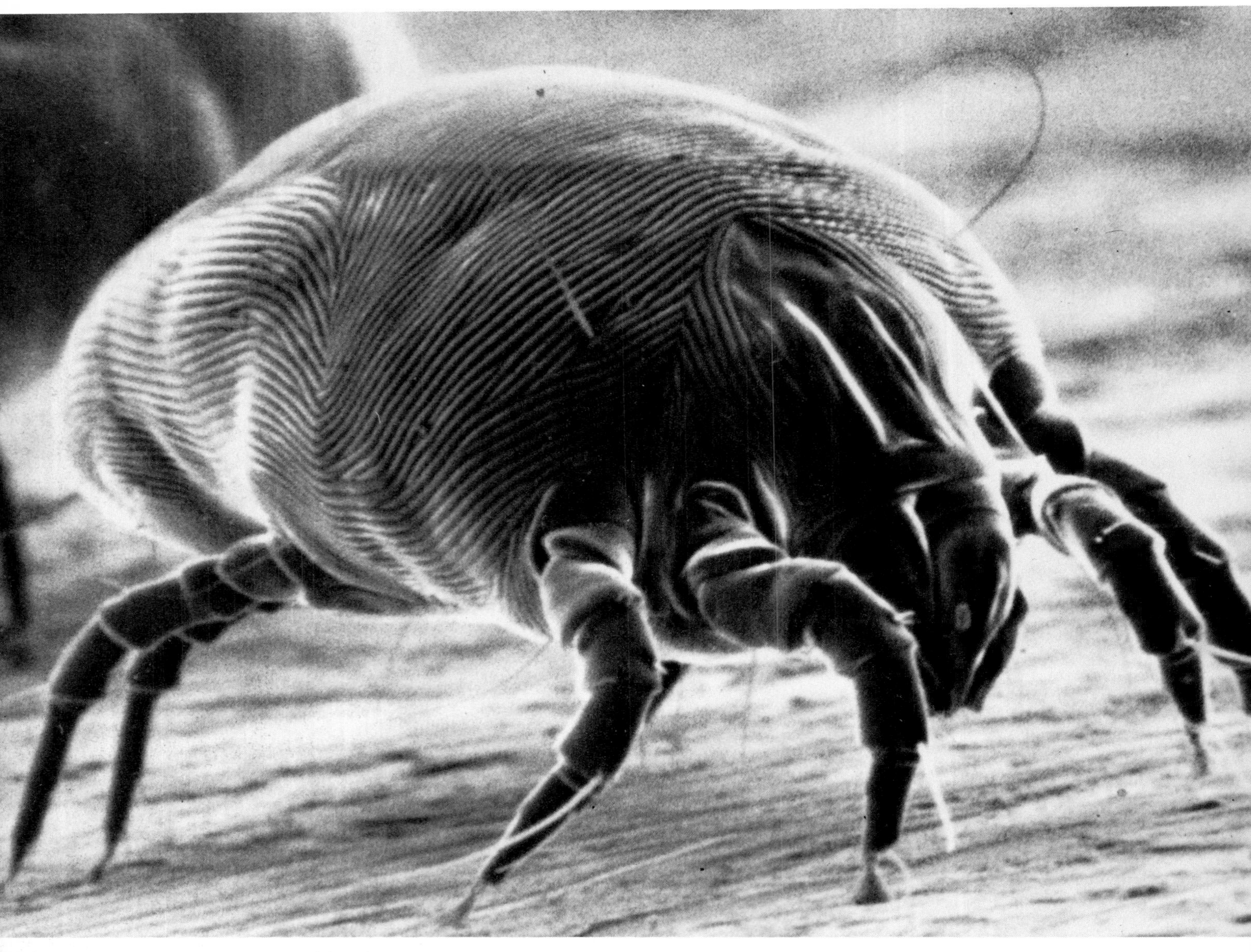

Why is house-dust extract, prepared by pharmaceutical houses, useful in testing for and treating house-dust allergies? Allergy specialists regularly use extracts of house dust to test allergic patients and to treat those found to be sensitive to this ubiquitous substance. But until recently allergists did not know what component of dust caused the allergic reaction. Now it seems that the agent responsible for the puffy eyes and sniffling noses of allergy victims is the almost invisible house mite, *Dermatophagoides farinae*. Scientists at Ohio State University tested samples of commercially prepared house-dust extract from many states and found that specimens from every major region contained the offending mite. The situation is one more example of the development of a specific treatment for a disease before it is known just how the disease is caused.

400 X—Dr. G. W. Wharton, Ohio State University

Another breed of mite, *Aceria nyssae*, lives inside minute galls about one millimeter in diameter that develop on the leaves of the black gum tree. In this photograph, the top of the gall has been whacked off with a razor blade to reveal the mite colony inside. Sometimes two to three hundred mites live in a gall hardly bigger than a pinhead.
80 X—Dr. Walter J. Humphreys, University of Georgia

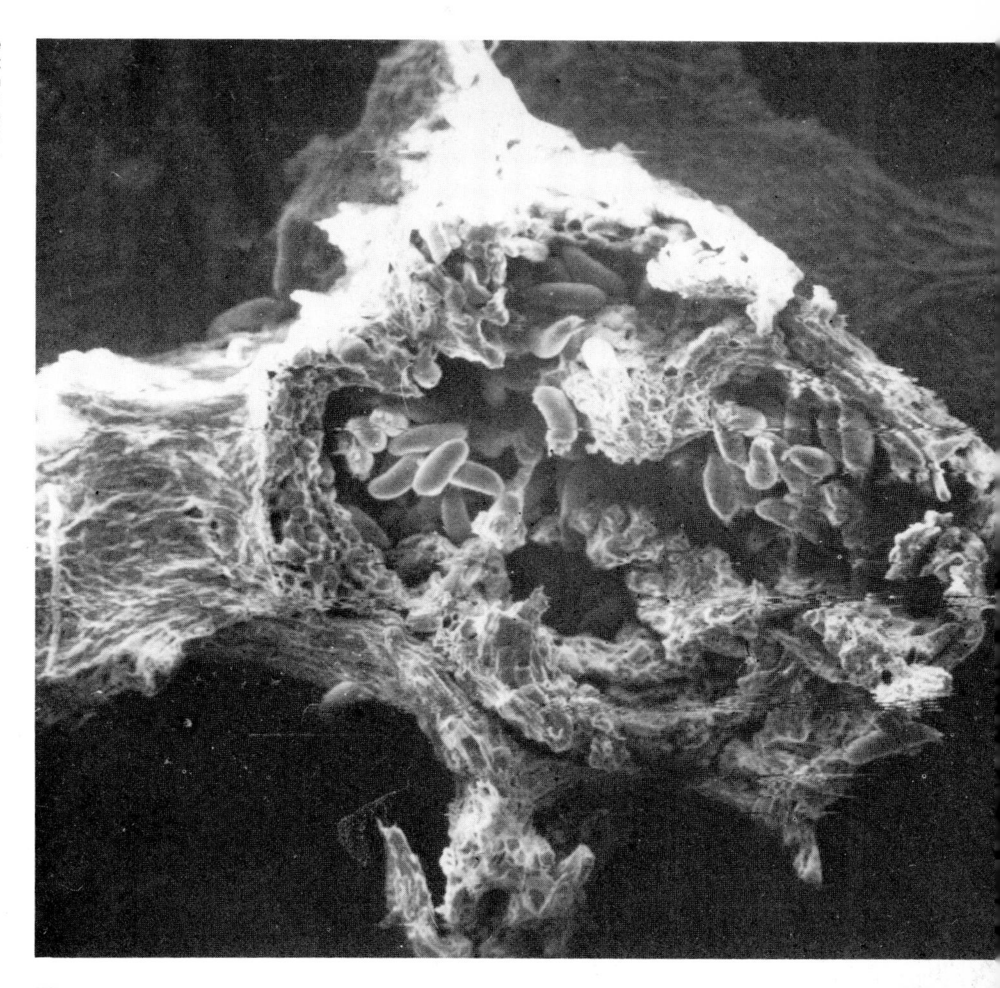

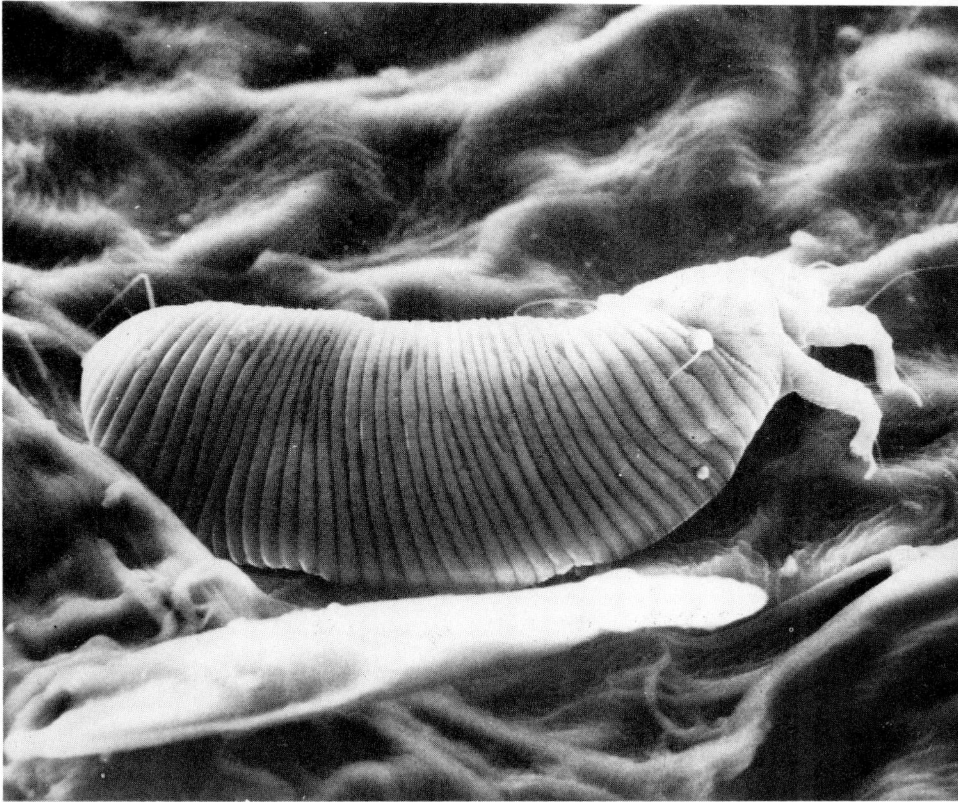

A single mite poses for his portrait.
1000 X—Dr. Walter J. Humphreys, University of Georgia

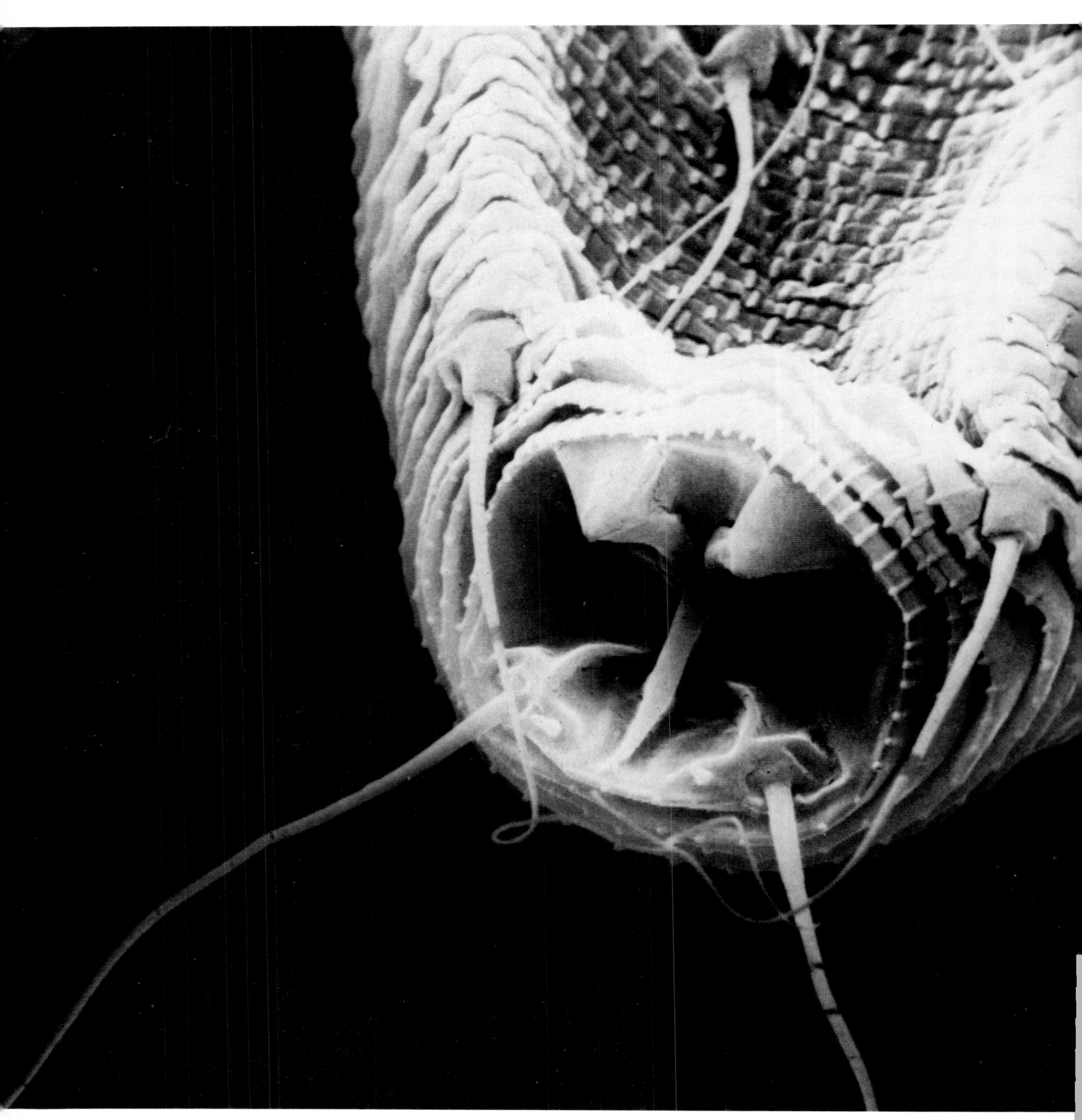

The scanning electron microscope reveals breathtaking detail in almost invisible biological specimens. This end view of the silver peach mite (*Aculus cornutus*) shows the sucker by which the animal attaches itself to a firm object to keep from being blown away by the wind.
4250 X—Dr. Walter J. Humphreys, University of Georgia

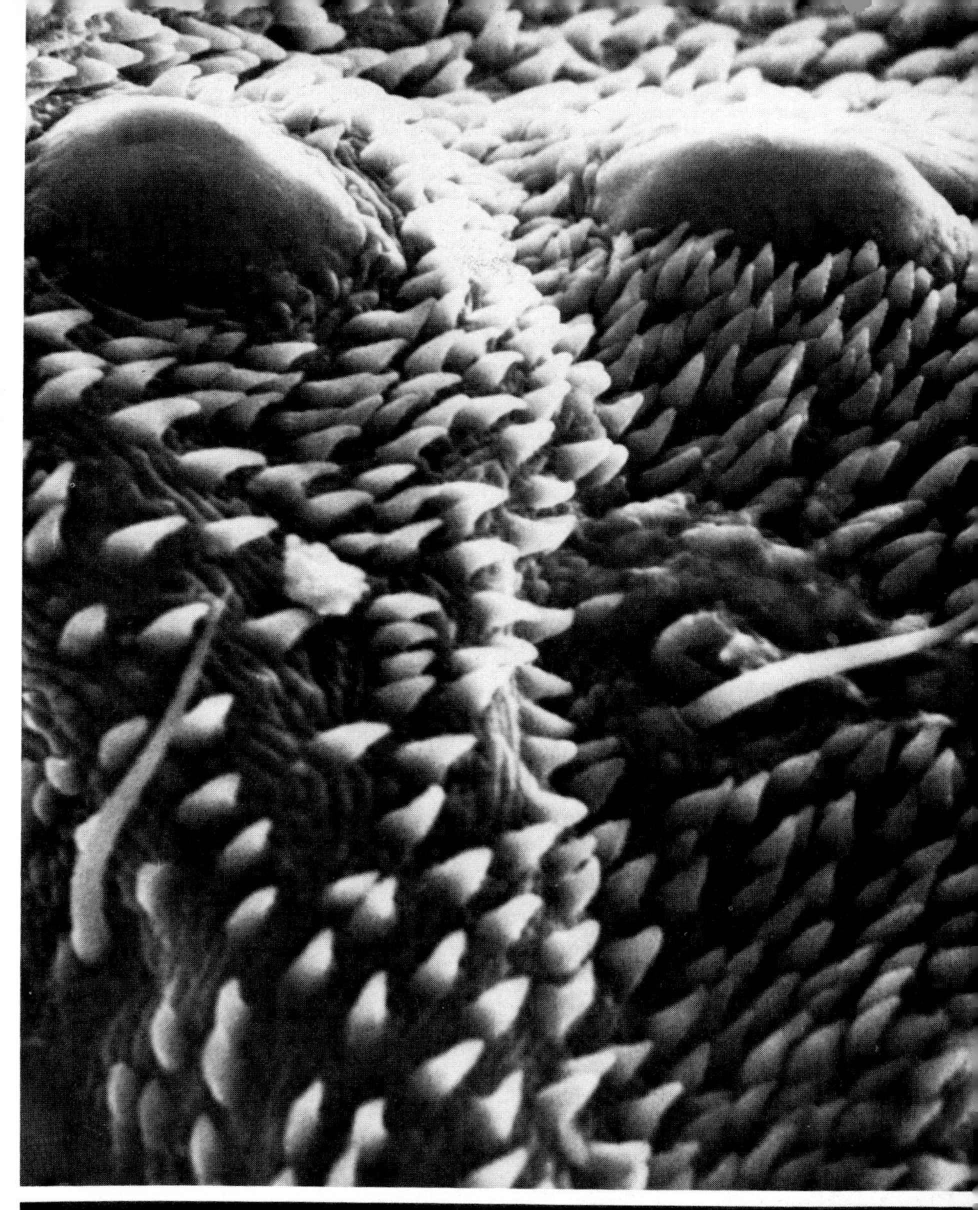

Those two mountains are the right pair of eyes of the water mite, *Tyrellia circularis*. The surrounding surface is covered with curious conelike structures.
650 X—Dr. G. W. Wharton, Ohio State University

Not a crowd of Ku Klux Klansmen seen from behind, but the strange, regular projections found on the back of the water mite, *Hydrophantes ruber*.
2400 X—Dr. G. W. Wharton, Ohio State University

The Gekko goes for a stroll up the inside of a bell jar. The tips of his toes are curling backward—an important clue in the mystery of how he clings to any surface. But its implication was not understood for years.
Conventional photo, three times life size—Dr. Joseph Gennaro, New York University

The Case of the Clinging Gekko

Small boys in Southeast Asia tie a string around the Gekko gecko, let it down from a second-story window, and jerk the hats from passersby. When not in captivity, the animal runs up and down panes of glass, across ceilings, and on other terrain where the footing is tricky, clinging effortlessly to any surface.

Some years ago, biologists examining the Gekko under a light microscope noted that the foot pads are filled with rows of fine fibers. They theorized that the lizard jams the fibers into tiny irregularities to hold on. But the Gekko's grip seemed too firm to be explained in that way.

The mystery was finally solved by Dr. Joseph F. Gennaro, of New York University, with the accompanying set of scanning electron microscope photographs.

In a closer view, the Gekko's foot is seen to be made up of ridges like the bottom of a ripple-soled shoe. These ridges, called lamellae, appear to be covered with a plush lining.
42 X—Dr. Joseph Gennaro, New York University

Magnified 1250 times, the plush turns into clusters of brushlike bristles. Earlier investigators had seen these structures and developed the theory that the Gekko got his grip by sticking the fibers into surface irregularities.
1250 X—Dr. Joseph Gennaro, New York University

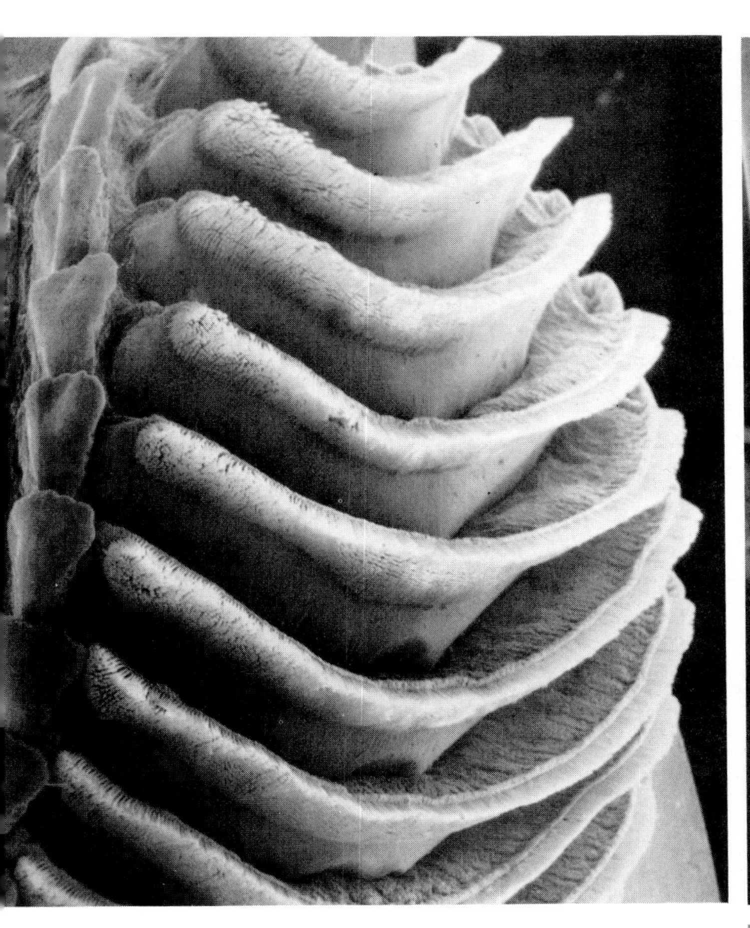

But no ordinary light microscope could reveal the tiny suction cups at the end of each hairlike fiber. Now investigators understand the Gekko's iron grip and also the reason his toes curl just before he lifts a foot (see first picture). With the tremendous clinging power furnished by the suction cups the Gekko wouldn't ordinarily have enough strength to pry his feet loose, so nature thoughtfully provided him with a toe-curling reflex that gently breaks the grip a bit at a time.
2650 X—Dr. Joseph Gennaro, New York University

CHAPTER 2

the world of the sea

Structures From the Sea

A major ridge gives this ostracode's shell great strength. As in a Gothic cathedral, buttresses support the ridge and impart even more strength and rigidity. With an economy often seen in nature, evolutionary forces have produced a series of holes in the structural members. These holes do not reduce strength, but lighten structure and cut down on the amount of raw material the creature must find for shell construction.
1500 X—Dr. Richard H. Benson, Smithsonian Institution

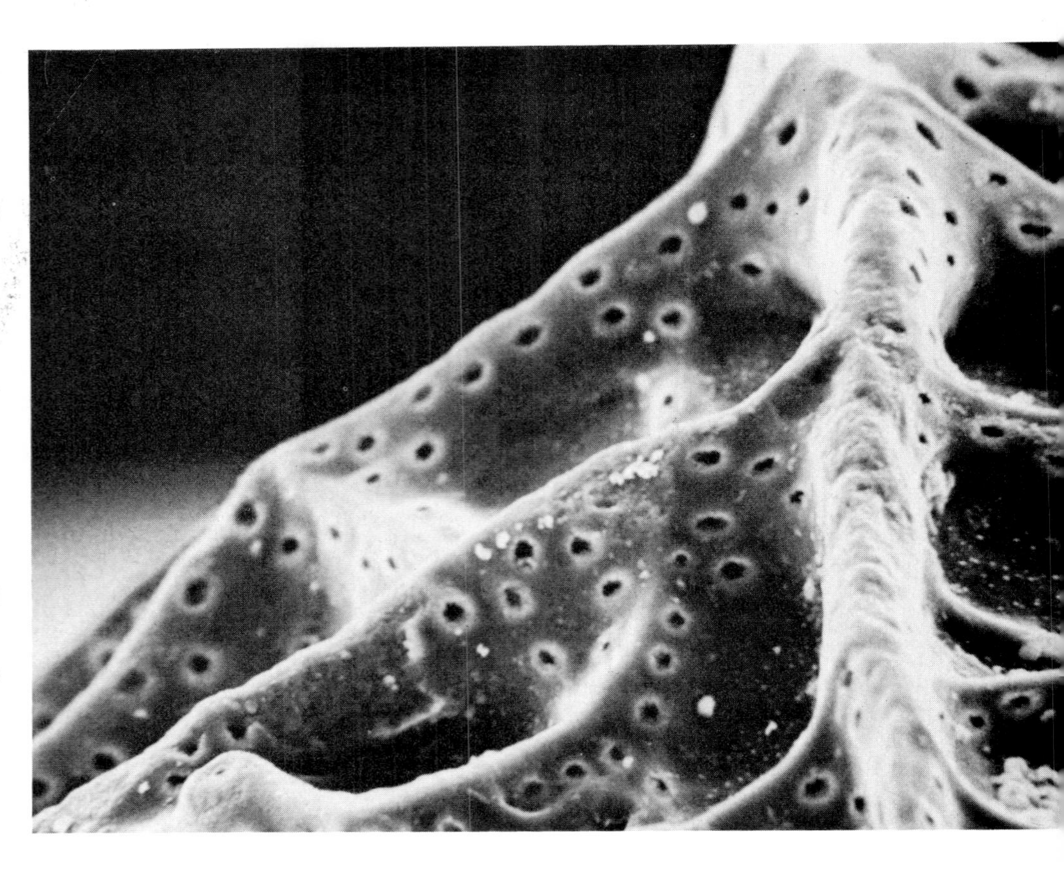

Nature, a talented architect, often finds incredibly beautiful solutions to straightforward engineering problems. This principle is illustrated perhaps nowhere more dramatically than in the shell design of the ostracode, a minute marine creature averaging a millimeter in length that lives in every type of aquatic environment from hot springs to the abyssal sea.

The ostracode's chief problem, in common with the rest of the animal and vegetable world, is survival. Since it is a crustacean, it has dealt with the matter of physical protection by developing a tough outer shell. Various members of the ostracode group have blossomed forth in an unimaginable variety of shell types, which make elegant use of such clearly identifiable engineering devices as trusses, beams, and buttresses.

Large enough to be seen with the naked eye, ostracodes had been examined in some detail with the light microscope, but the limitations of that instrument are severe. The scanning electron microscope, on the other hand, has made even the most minute features of the ostracode carapace visible in striking detail, bringing about what one investigator called a "revolutionary climate" in micropaleontology.

An external truss system, similar to the ones used by bridge engineers, furnishes support for the protective shell inside.
600 X—Dr. Richard H. Benson, Smithsonian Institution

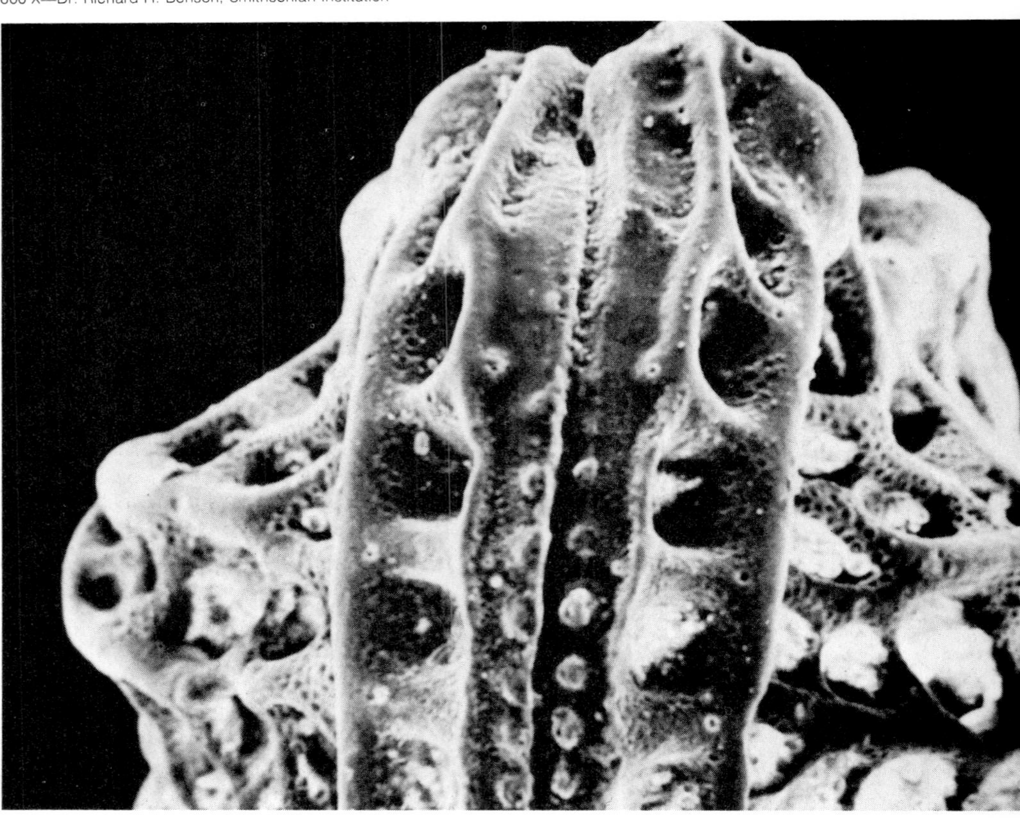

A simple smooth shell would have some strength. But the three arched beams with their supporting buttresses make this creature's shell many times stronger with only a small amount of additional material.
110 X—Dr. Richard H. Benson, Smithsonian Institution

Local stresses in this shell are distributed over a wide area by a network of spiny beams.
750 X—Dr. Richard H. Benson, Smithsonian Institution

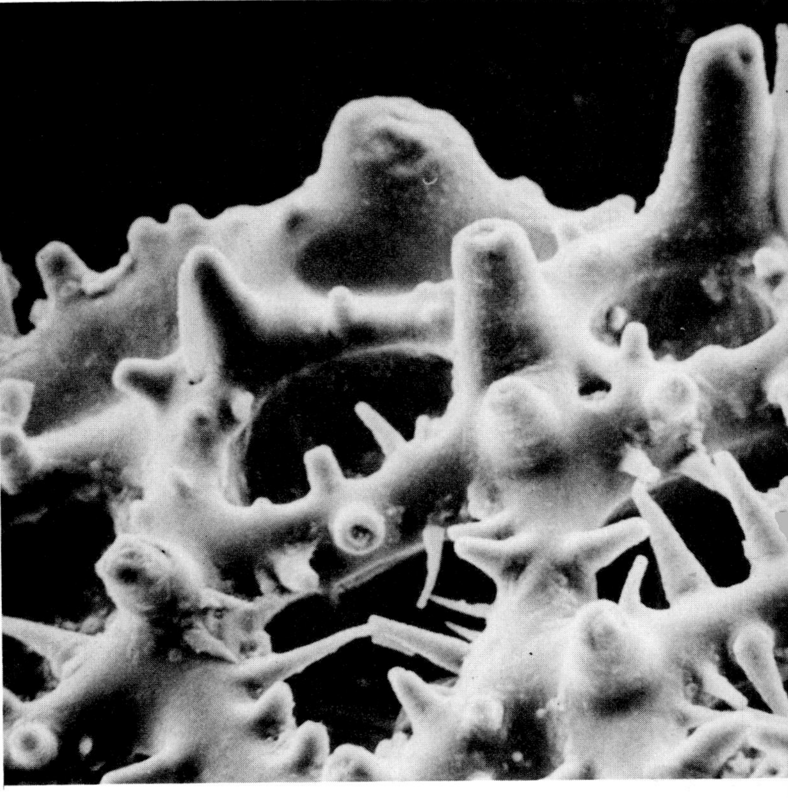

Single elements of this fail-safe, honeycomblike truss system can be destroyed without seriously affecting overall strength. Minute holes reduce weight.
1100 X—Dr. Richard H. Benson, Smithsonian Institution

The irregular shape of this creature calls for special structural strengthening. *Paracytheridea* developed it in the form of delicate, veinlike supporting ridges.
175 X—Dr. Richard H. Benson, Smithsonian Institution

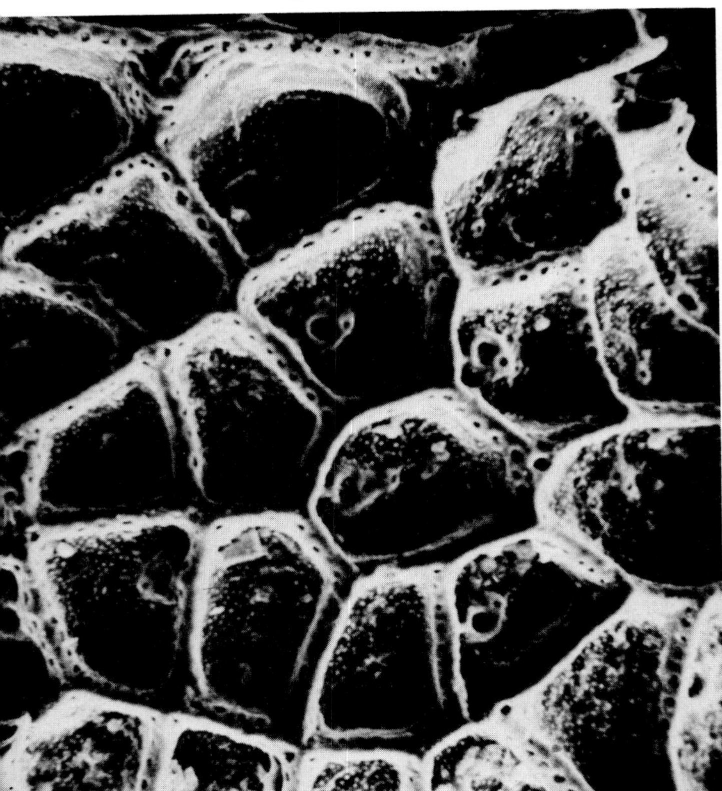

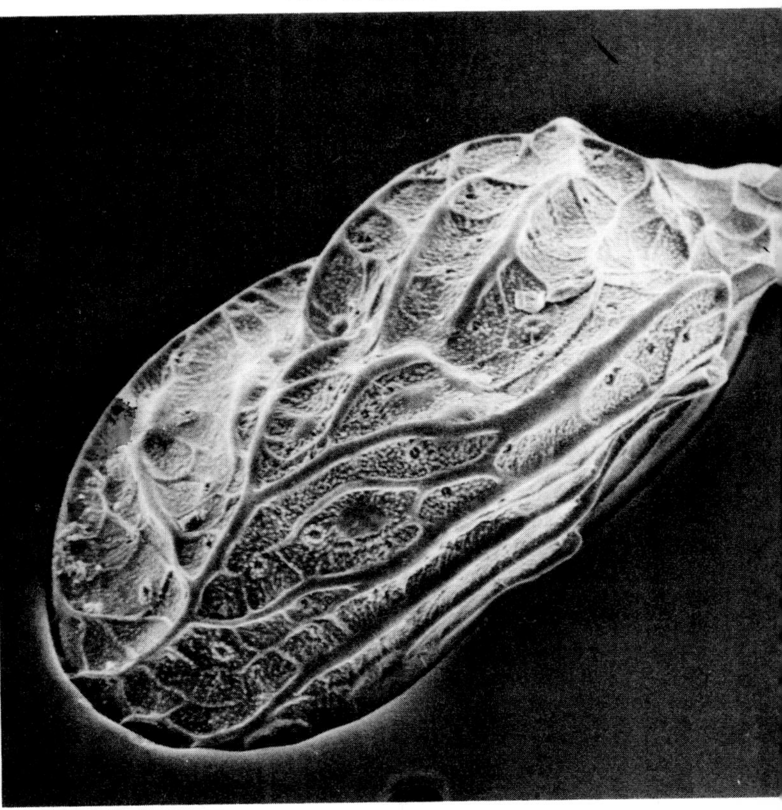

The girder-like formation of ribs into a highly organized bridgelike system of trusses marks *Jugosocytheresis*, a delicately sculptured ostracode from tropical seas. A multitude of structure-lightening holes add to the delicate effect.
200 X—Dr. Richard H. Benson, Smithsonian Institution

Over the eons, *Reticulocythesis* evolved in the Indian Ocean. Today, bold supporting ridges eliminate the need for the smaller intermediate structures its ancestors once had, so they are slowly disappearing. The pits formed by the receding structures form the lovely and unusual cloverleaf pattern.
220 X—Dr. Richard H. Benson, Smithsonian Institution

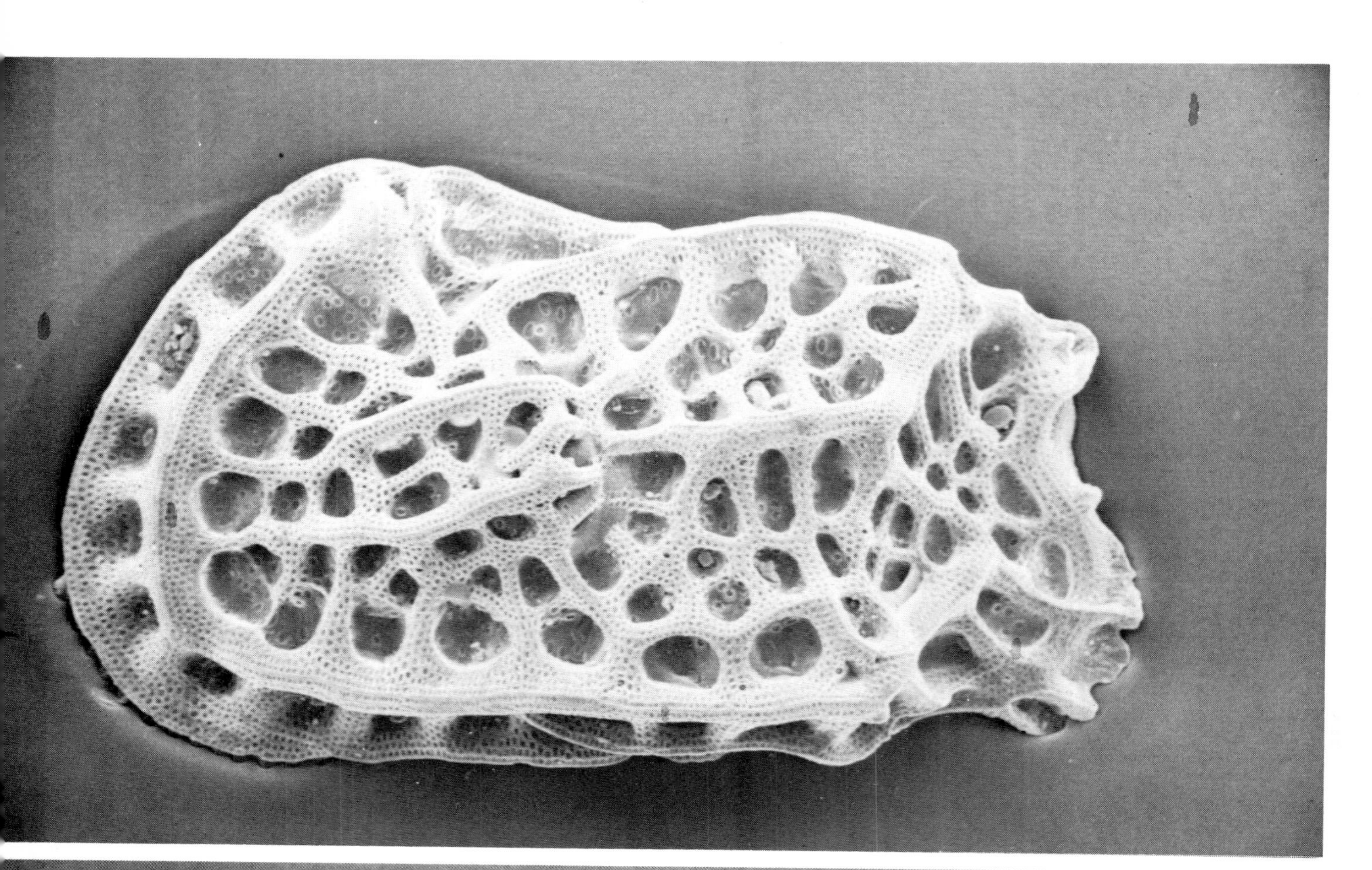

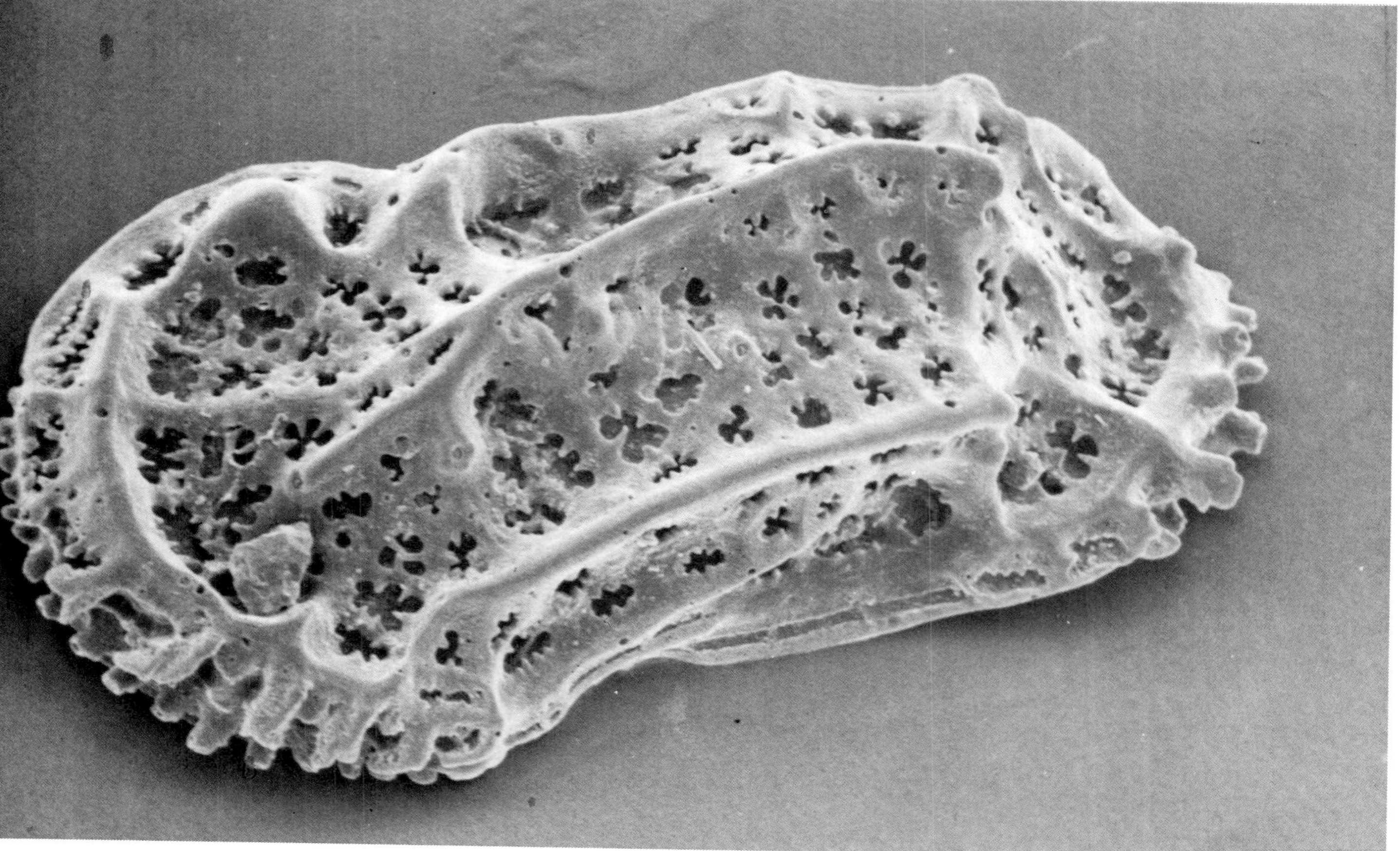

Tremendously diverse conditions of temperature, wave action, and other factors prompted these lovely solutions to the basic problem of survival.

185 X—Dr. Richard H. Benson, Smithsonian Institution

135 X—Dr. Richard H. Benson, Smithsonian Institution

135 X—Dr. Richard H. Benson, Smithsonian Institution

150 X—Dr. Richard H. Benson, Smithsonian Institution

Fabulous Fossils

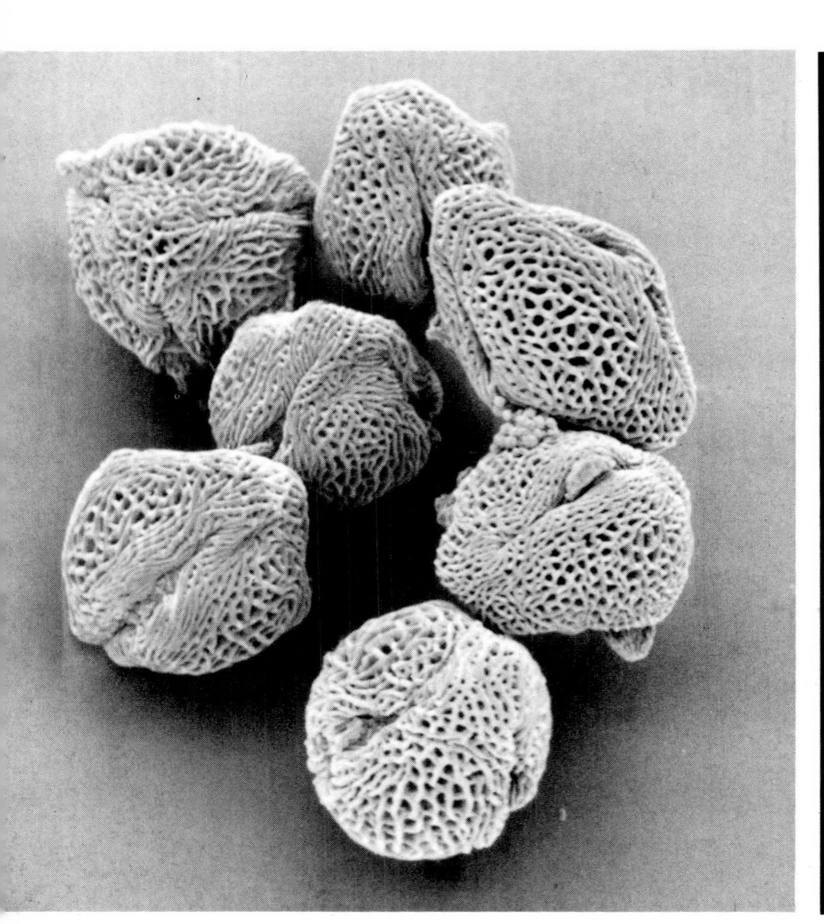

Pollen
1450 X—Esso Production Research Company

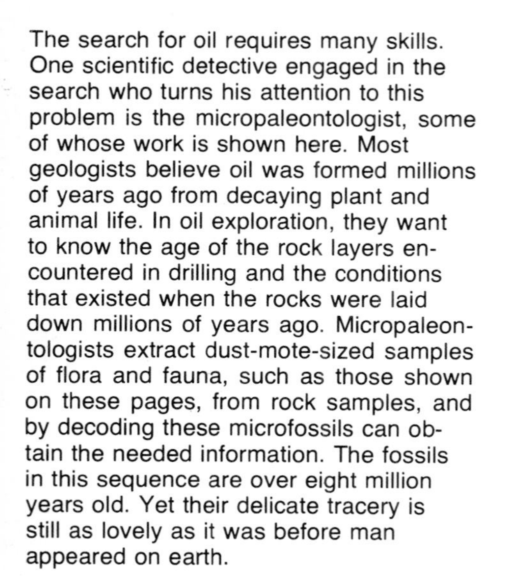

The search for oil requires many skills. One scientific detective engaged in the search who turns his attention to this problem is the micropaleontologist, some of whose work is shown here. Most geologists believe oil was formed millions of years ago from decaying plant and animal life. In oil exploration, they want to know the age of the rock layers encountered in drilling and the conditions that existed when the rocks were laid down millions of years ago. Micropaleontologists extract dust-mote-sized samples of flora and fauna, such as those shown on these pages, from rock samples, and by decoding these microfossils can obtain the needed information. The fossils in this sequence are over eight million years old. Yet their delicate tracery is still as lovely as it was before man appeared on earth.

Pollen
3250 X—Esso Production Research Company

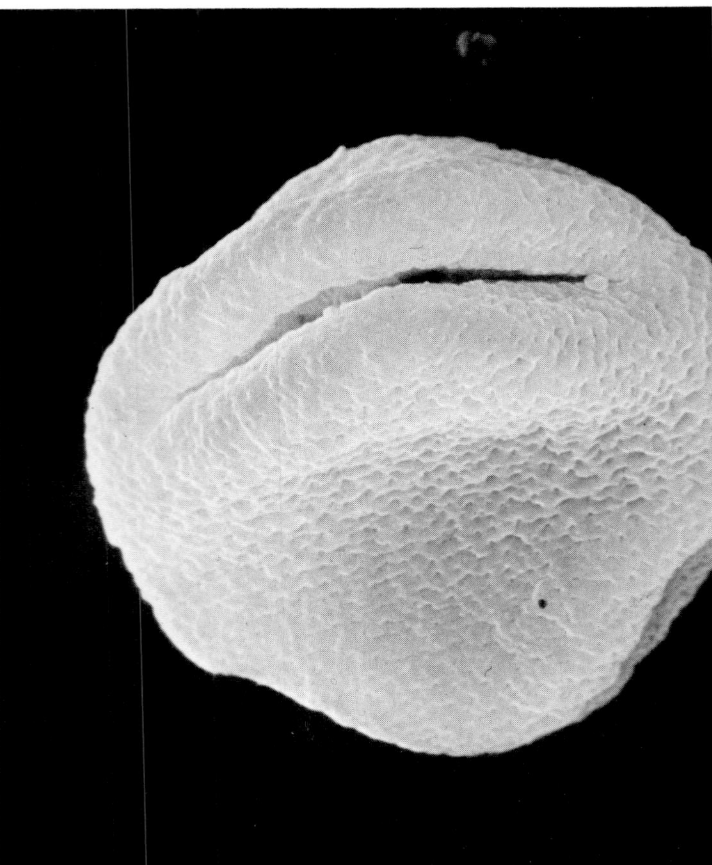

Foraminifer
185 X—Esso Production Research Company

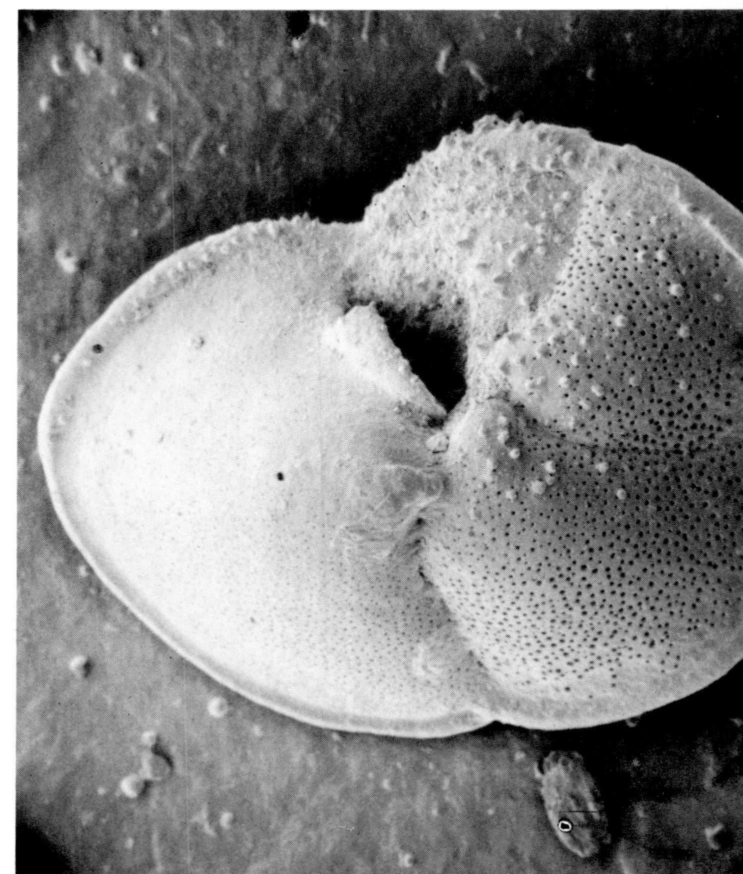

Diatom
1500 X—Esso Production Research Company

Chitinozoan
650 X—Esso Production Research Company

Dinoflagellate
1250 X—Esso Production Research Company

Discoaster
2500 X—Esso Production Research Company

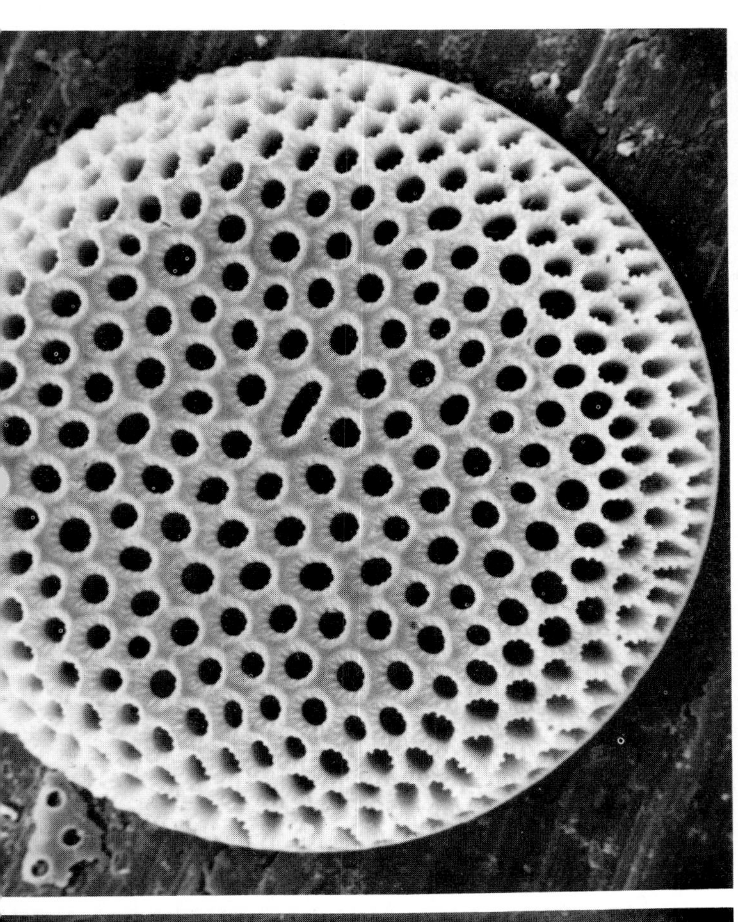

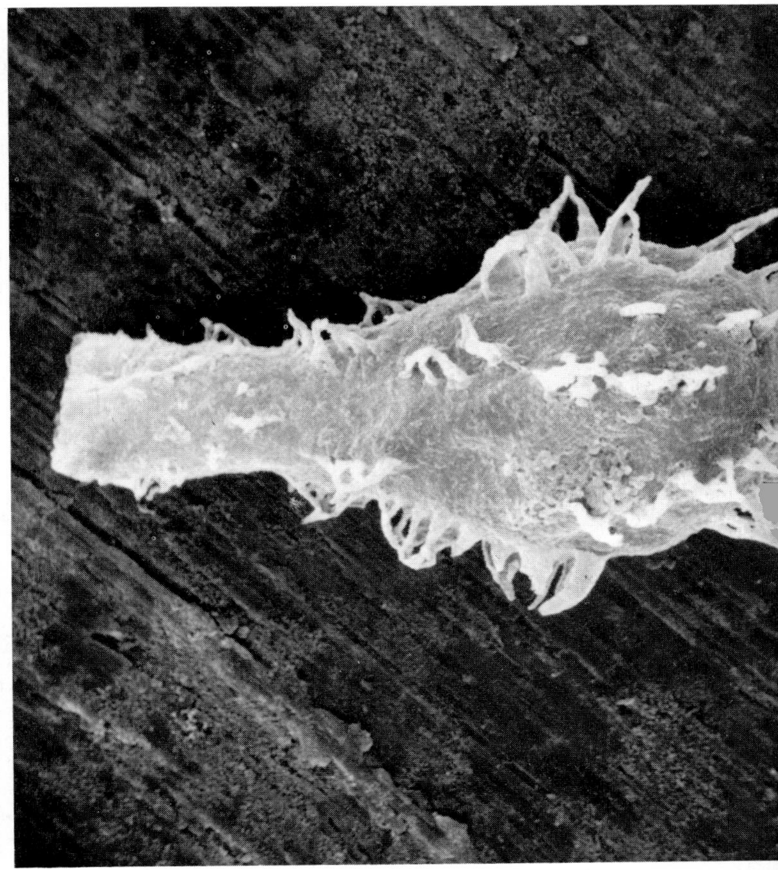

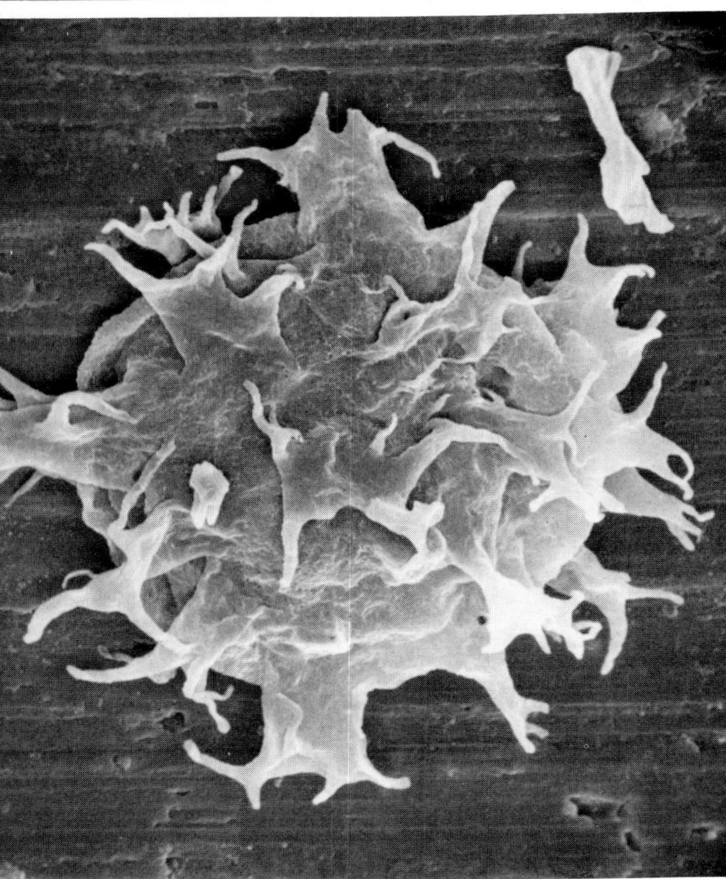

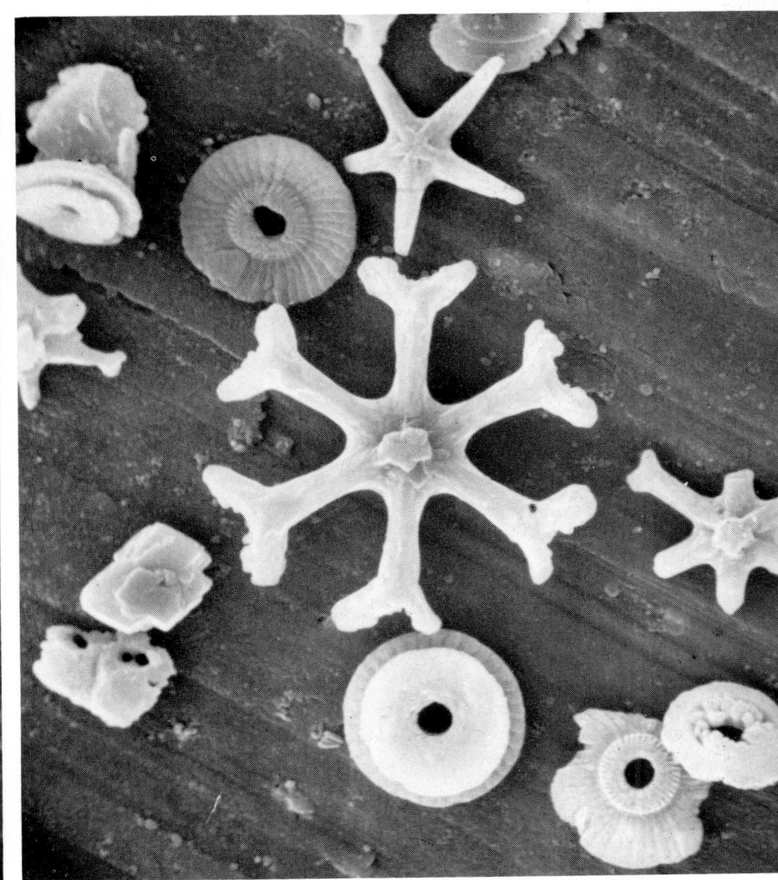

Diatoms

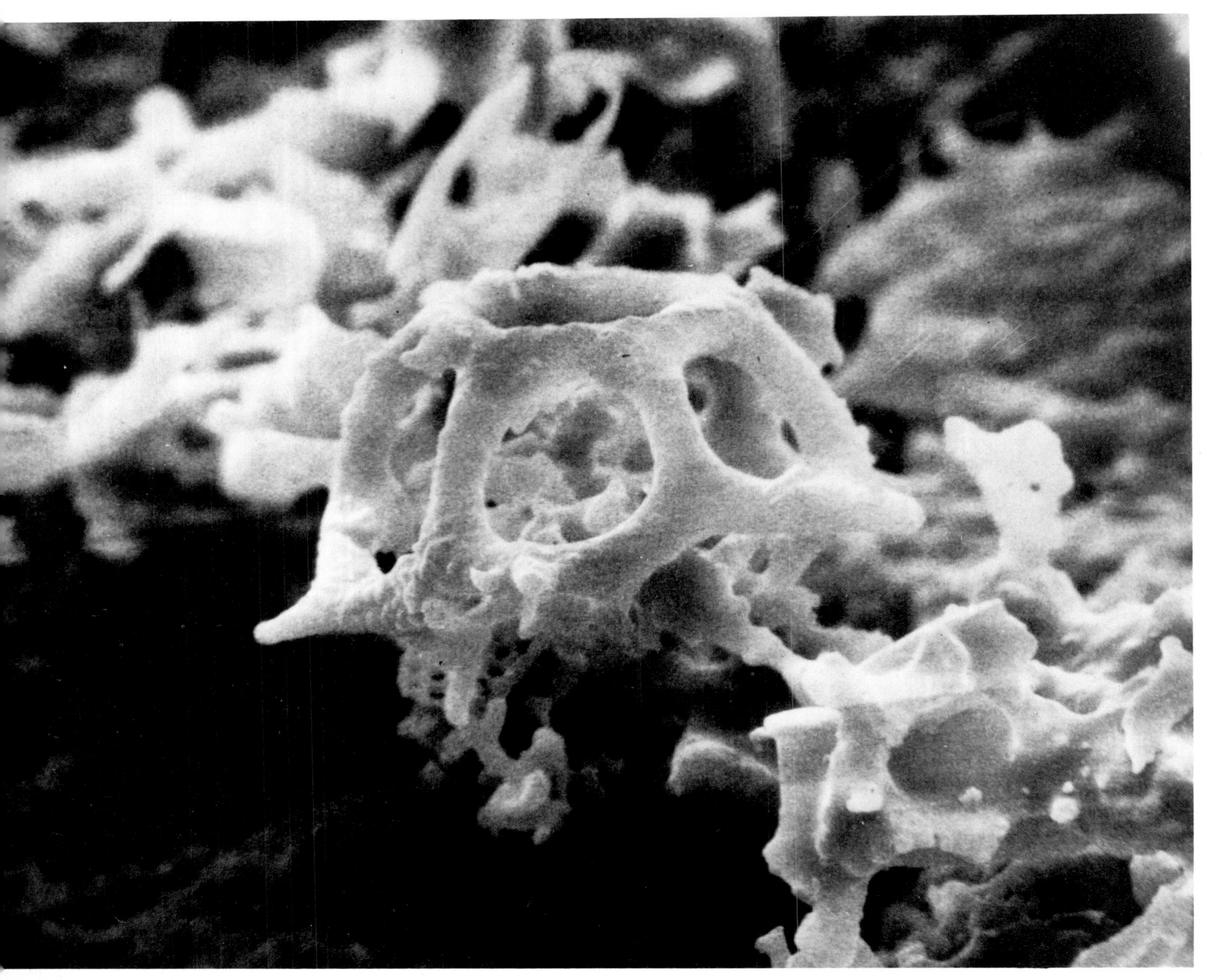

12,000 X—David Sarnoff Research Center, Radio Corporation of America

Diatoms are one-celled brownish-green sea plants enclosed in transparent silica shells of great variety and beauty. When the plants die, the shells sink to the floor of the sea. Here, two particularly lovely examples: a miniature crown and a startling barrel-like structure.

3400 X—Miss B. Farber, Johnson and Johnson Research

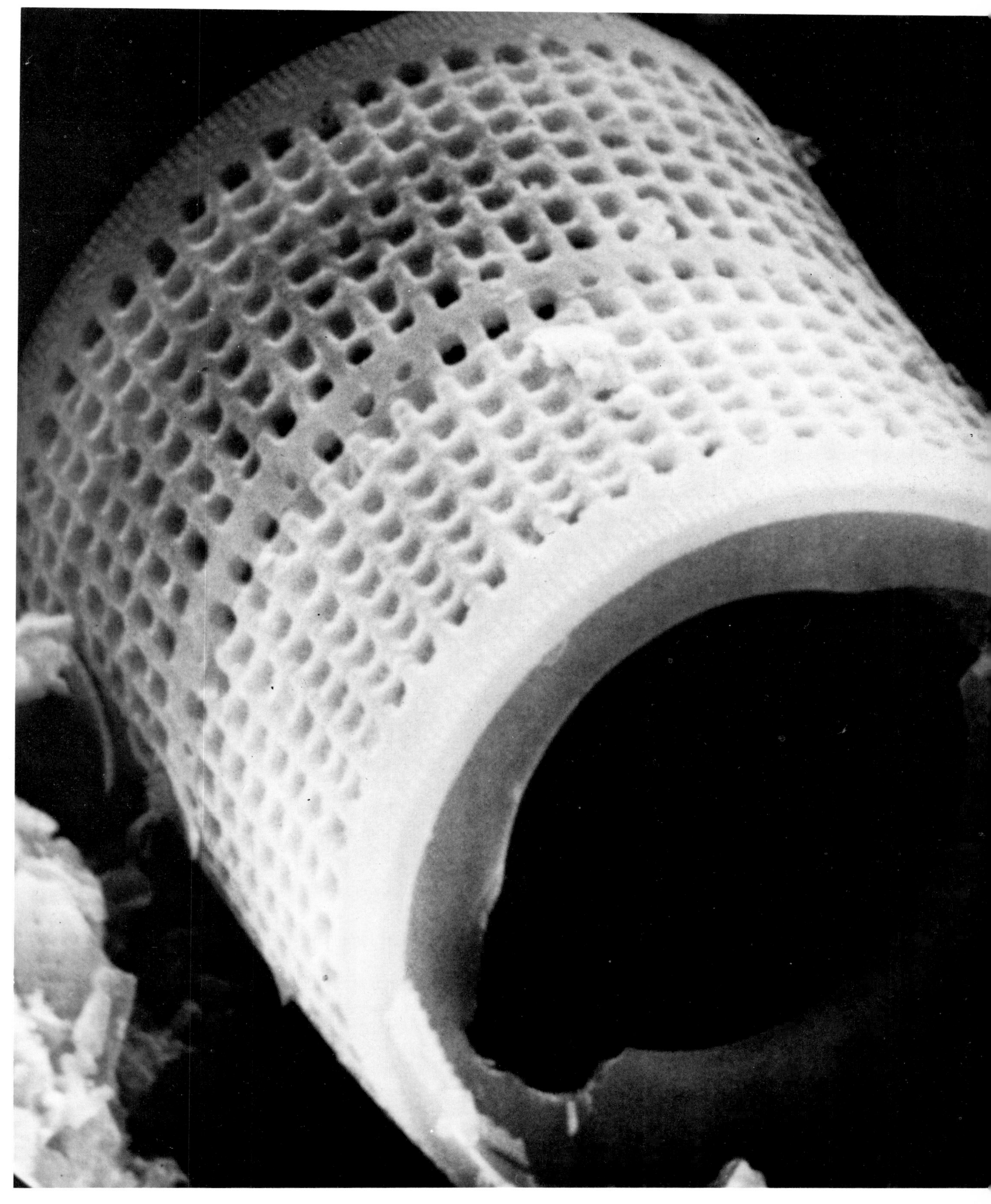

Mother-of-Pearl

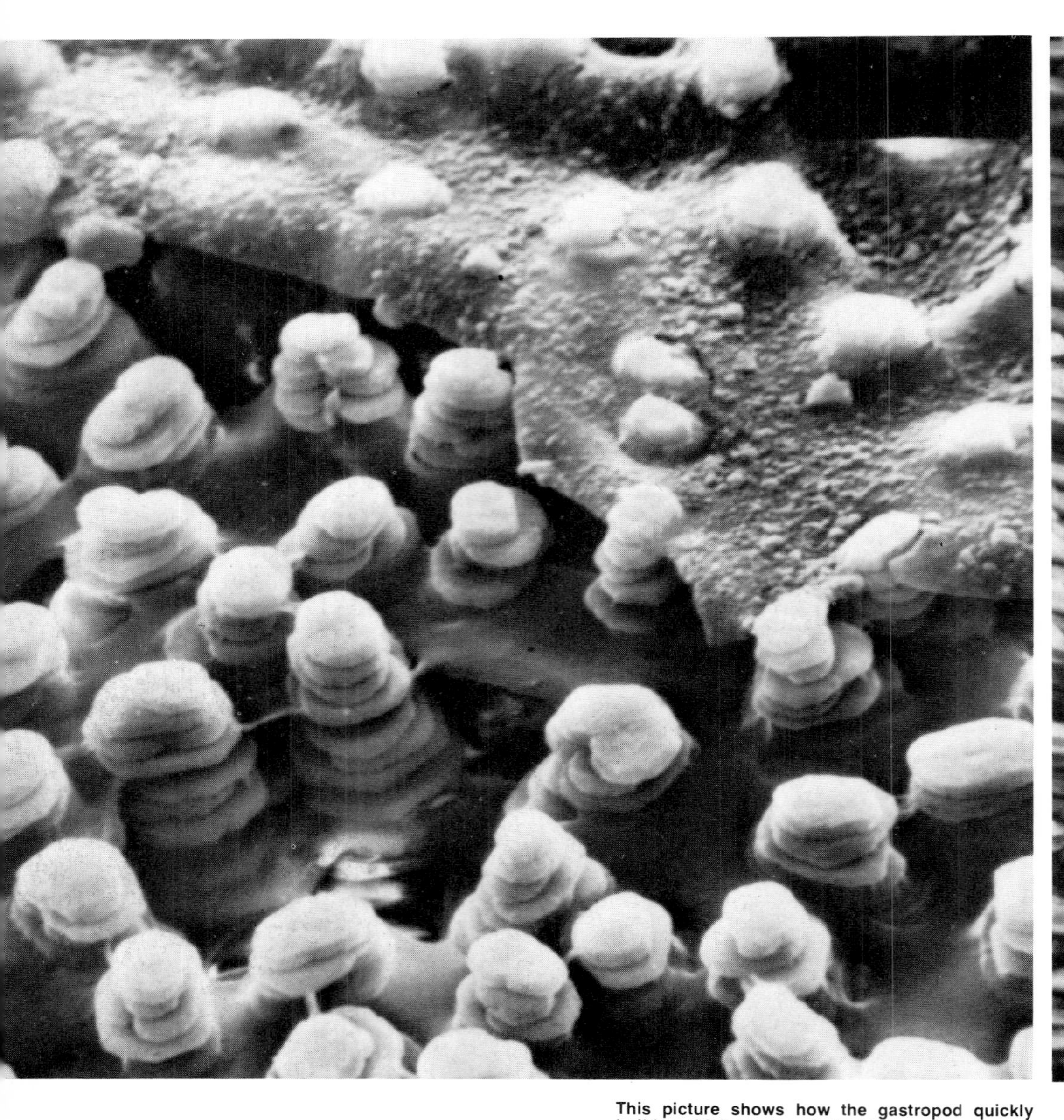

This picture shows how the gastropod quickly builds a thick section of shell in a series of cone-shaped stacks. New crystals form at the tops of the stacks. Meanwhile, older crystals below continue to grow outward. Ultimately, the crystals in adjacent stacks meet, forming a solid layer.

1600 X—Dr. Sherwood W. Wise, Jr., Florida State University

The evolutionary thrust of nature develops many ingenious solutions to the problems of existence and growth. Two elegant approaches are the strikingly different but equally effective methods of generating that lustrous, iridescent material, mother-of-pearl, which are used by two species of mollusks. The two families are gastropods (snails, periwinkles, conchs, and other creatures usually having spiral or cup-shaped shells) and pelecypods (oysters, mussels, clams, and other bivalves). The gastropod shell grows outward, continually building a more complex spiral or other shape. The growth zone is only a few millimeters wide, so new material must be laid down to the total ultimate thickness of the mother-of-pearl layer in a very short time.

regular structure of the crystals is shown in cross section of gastropod mother-of-pearl ained by breaking a shell. The platelike crystals ning the mother-of-pearl line up in precise verti- stacks.
X—Dr. Sherwood Wise, Jr., Florida State University

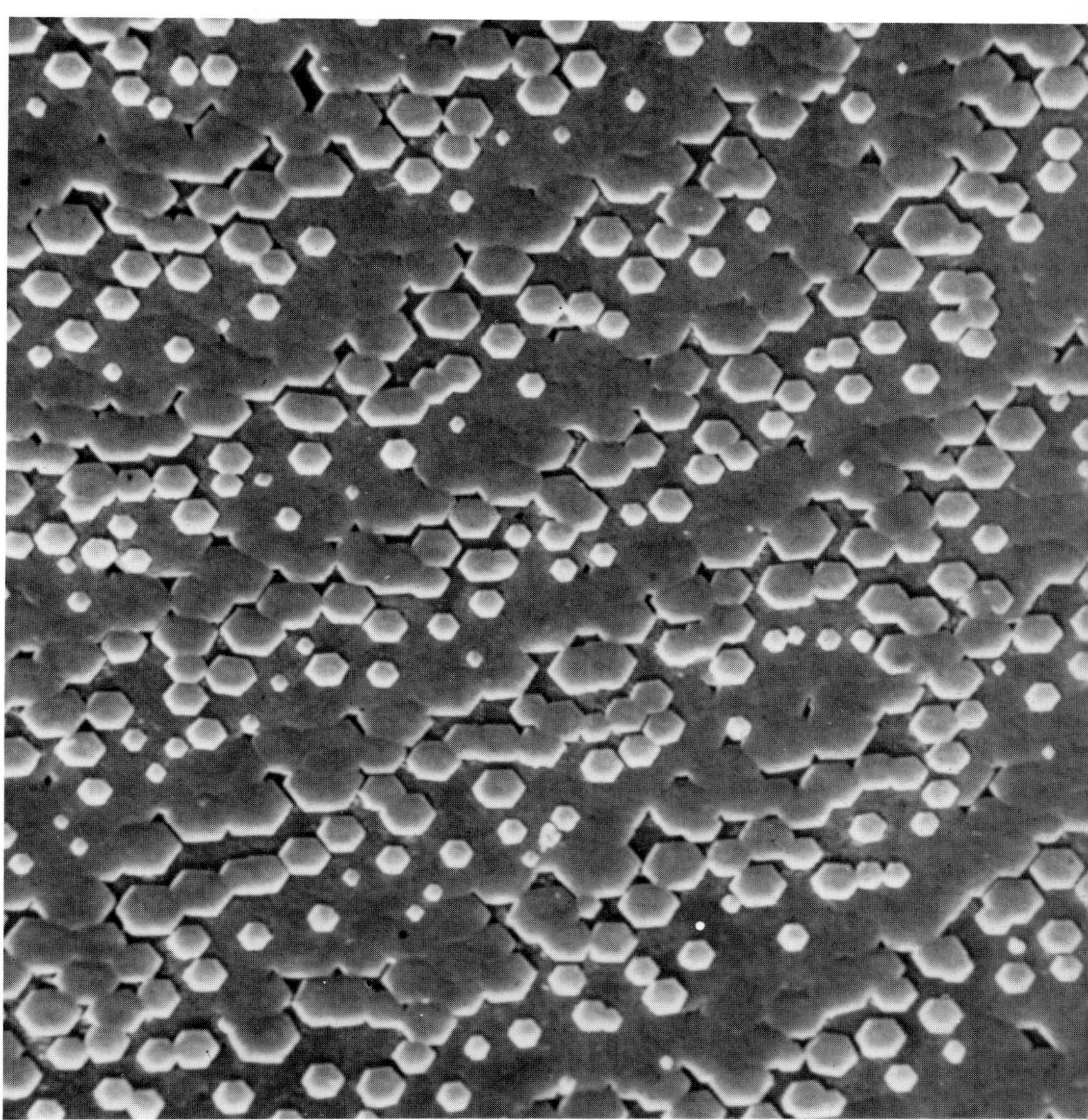

The pelecypod begins life with a very thin layer of mother-of-pearl inside its shell and keeps adding layer after layer for as long as it lives. This picture shows the lovely pattern of hexagonal crystals that develops in a steplike sequence as new crystals form at the edges of each step. Thus each family—gastropod and pelecypod—has developed the method of mother-of-pearl formation best suited to its shape and growth pattern.
1600 X—Dr. Sherwood Wise, Jr., Florida State University

Sea Urchin

The scanning electron microscope reveals the secret of the amazing strength of the sea urchin's spines. Each spine grows in the form of a single crystal in a many-windowed pattern. If a crack develops, it quickly runs into a window, which stops it, thus protecting the overall structure. ►
570 X—Dr. Jon Weber, The Pennsylvania State University

Sea urchin
Conventional photo, slightly larger than life size—Dr. Jon Weber, The Pennsylvania State University

It has long been known that the sea urchin's spines—the long, cylindrical structures attached to the animal's shell —are exceedingly light and strong. The animal can move them rapidly to ward off attack. Recent tests have shown that the spines are stronger for their weight than brick or concrete.

A closer look reveals arches which suggest a Henry Moore sculpture. Investigators theorize that it may be possible to build lighter, stronger structures using the principles employed in the sea urchin's shell. ►
7000 X—Dr. Jon Weber, The Pennsylvania State University

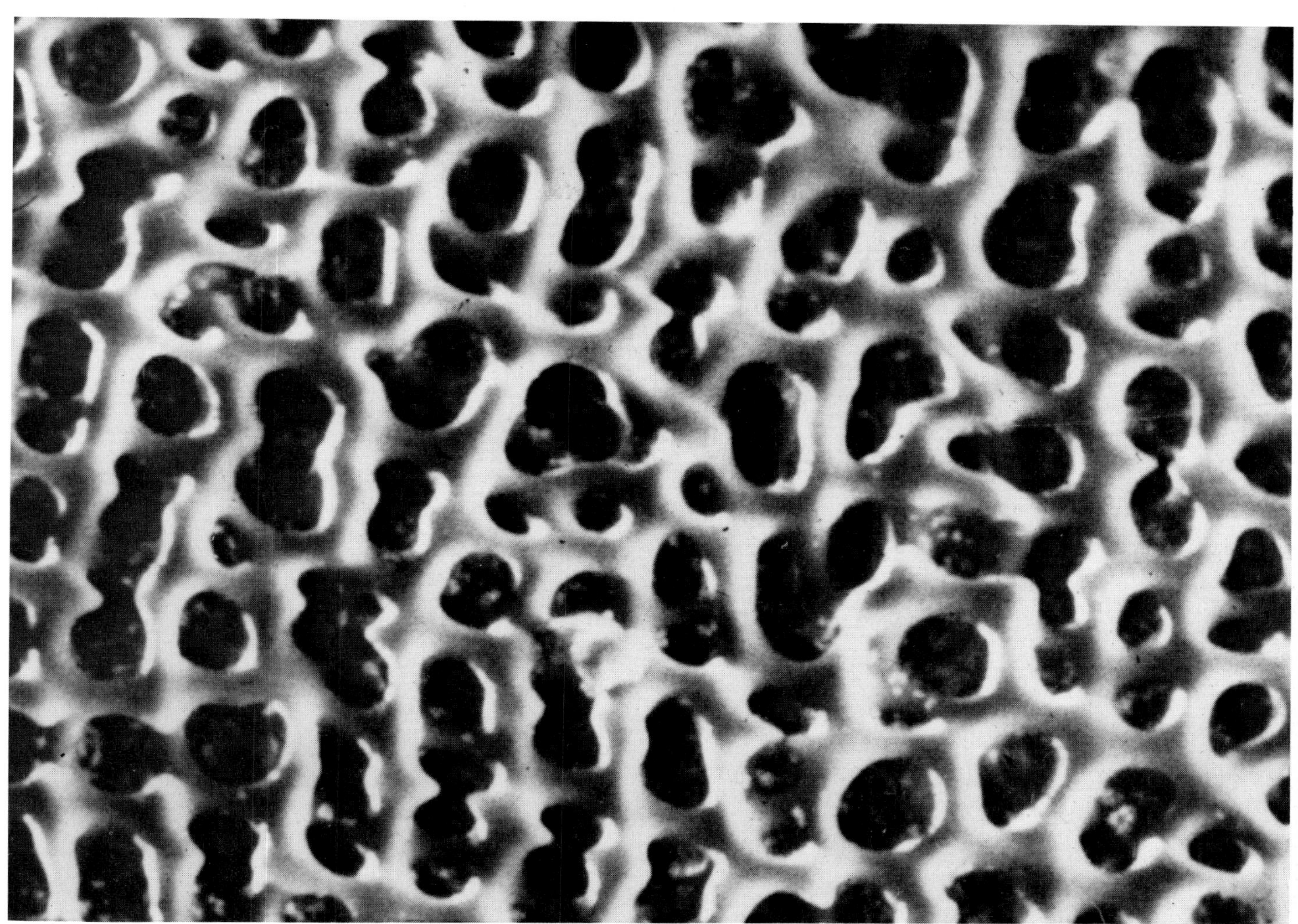

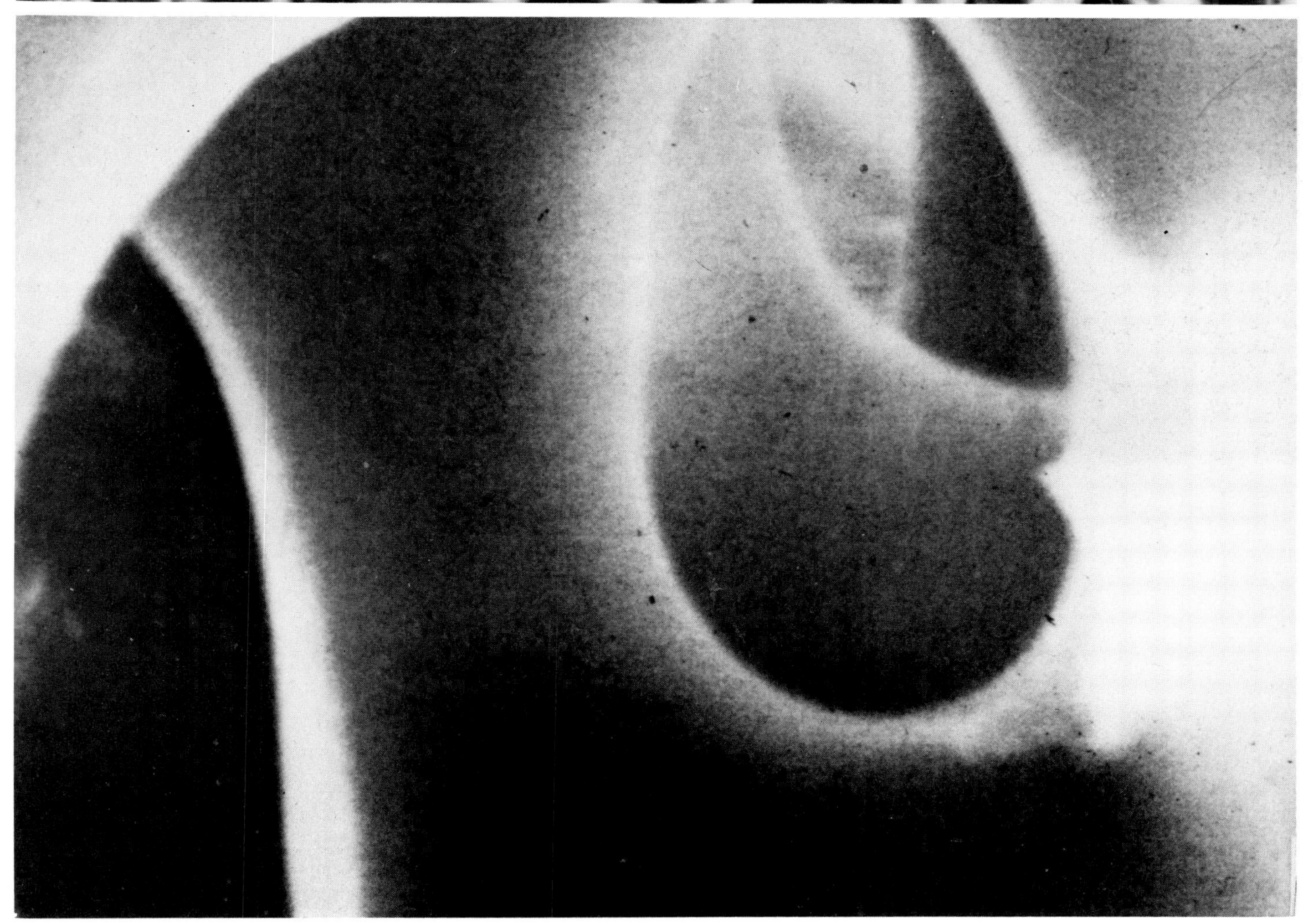

CHAPTER 3

around the house

Salt
100 X—Johnson Wax Photo

Pepper
100 X—Johnson Wax Photo

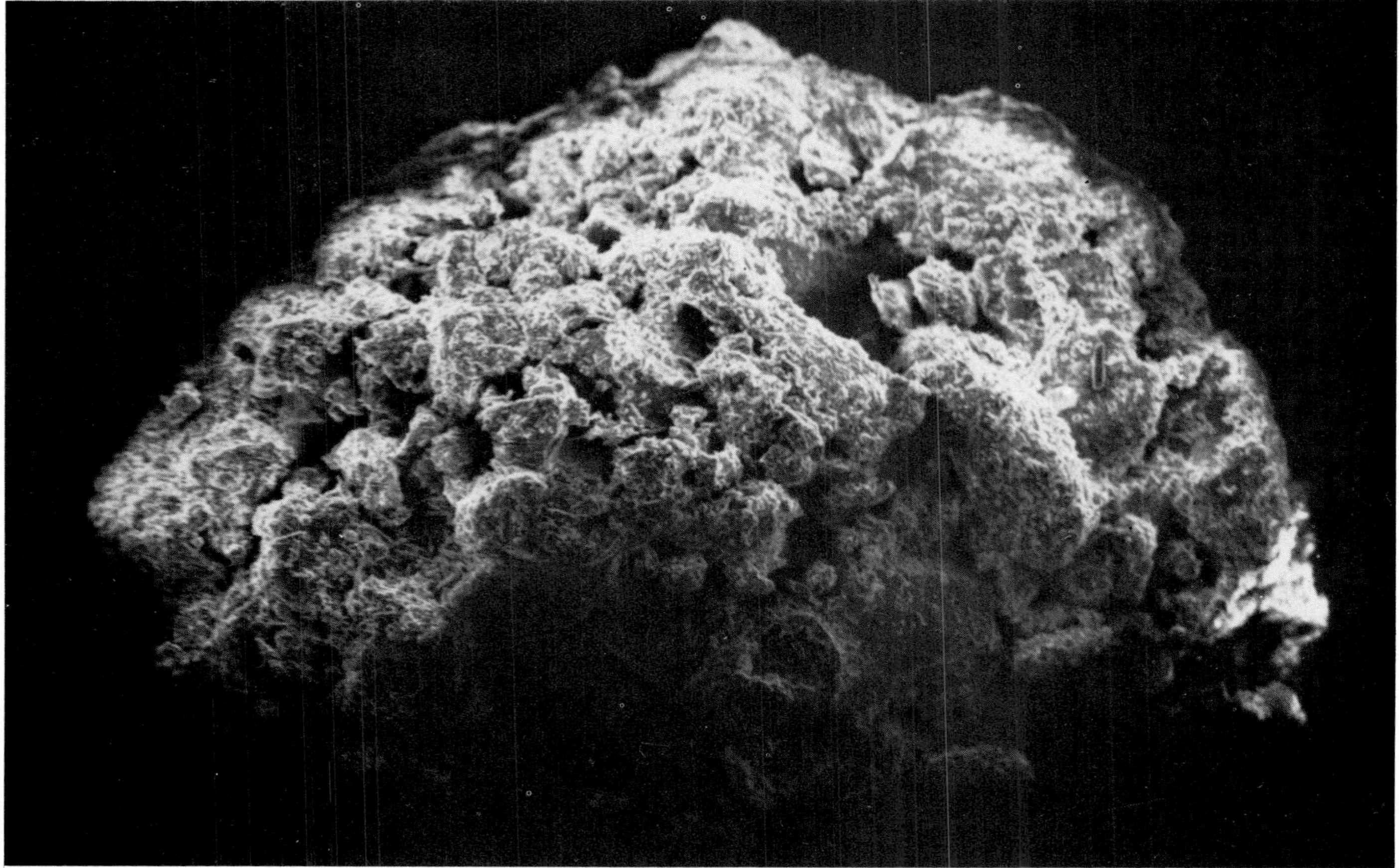

Ball-point pen
168 X—Johnson Wax Photo

Ball-point pen
1500 X—Dr. Emil Bernstein, Gillette Research Institute

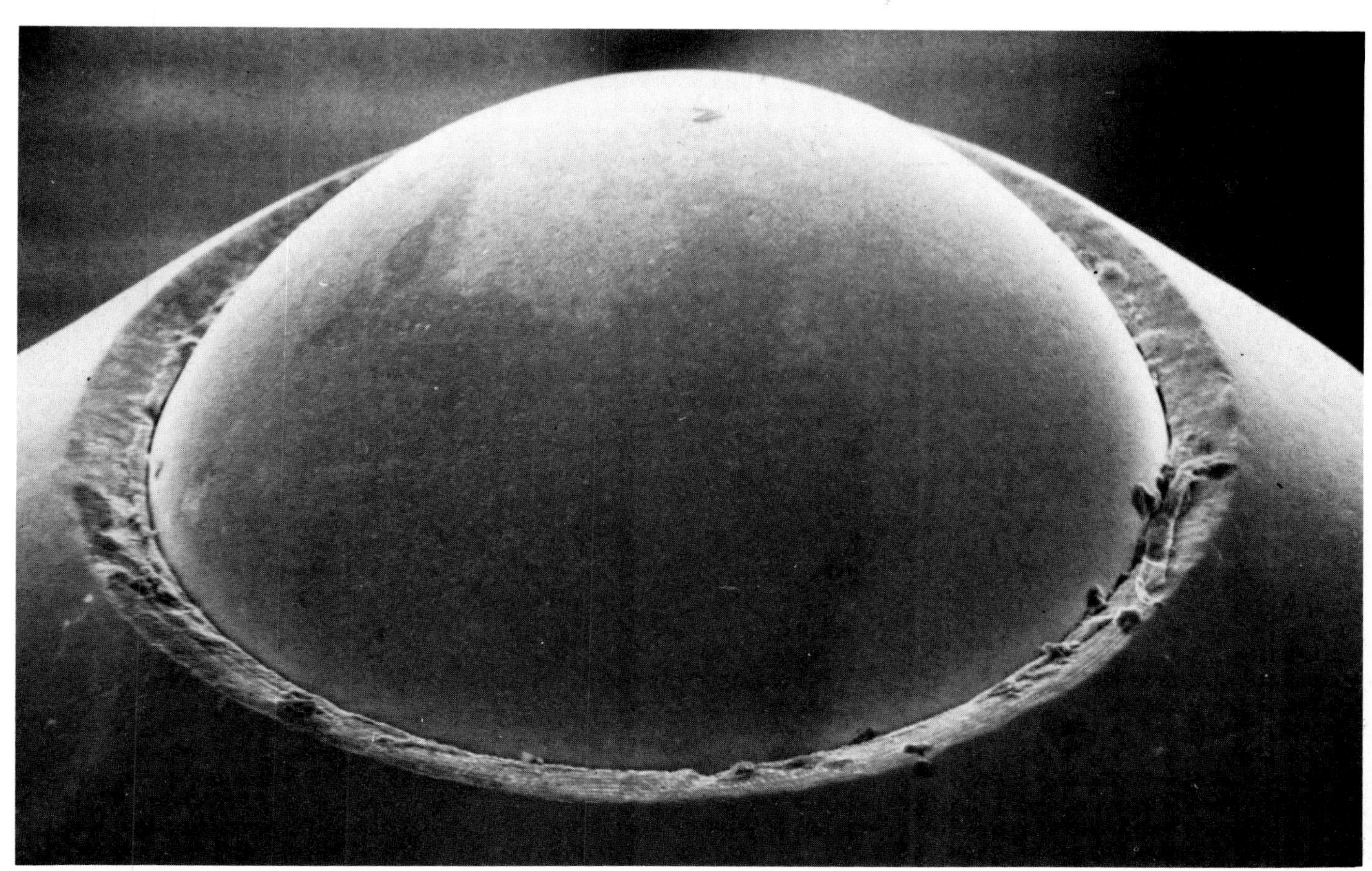

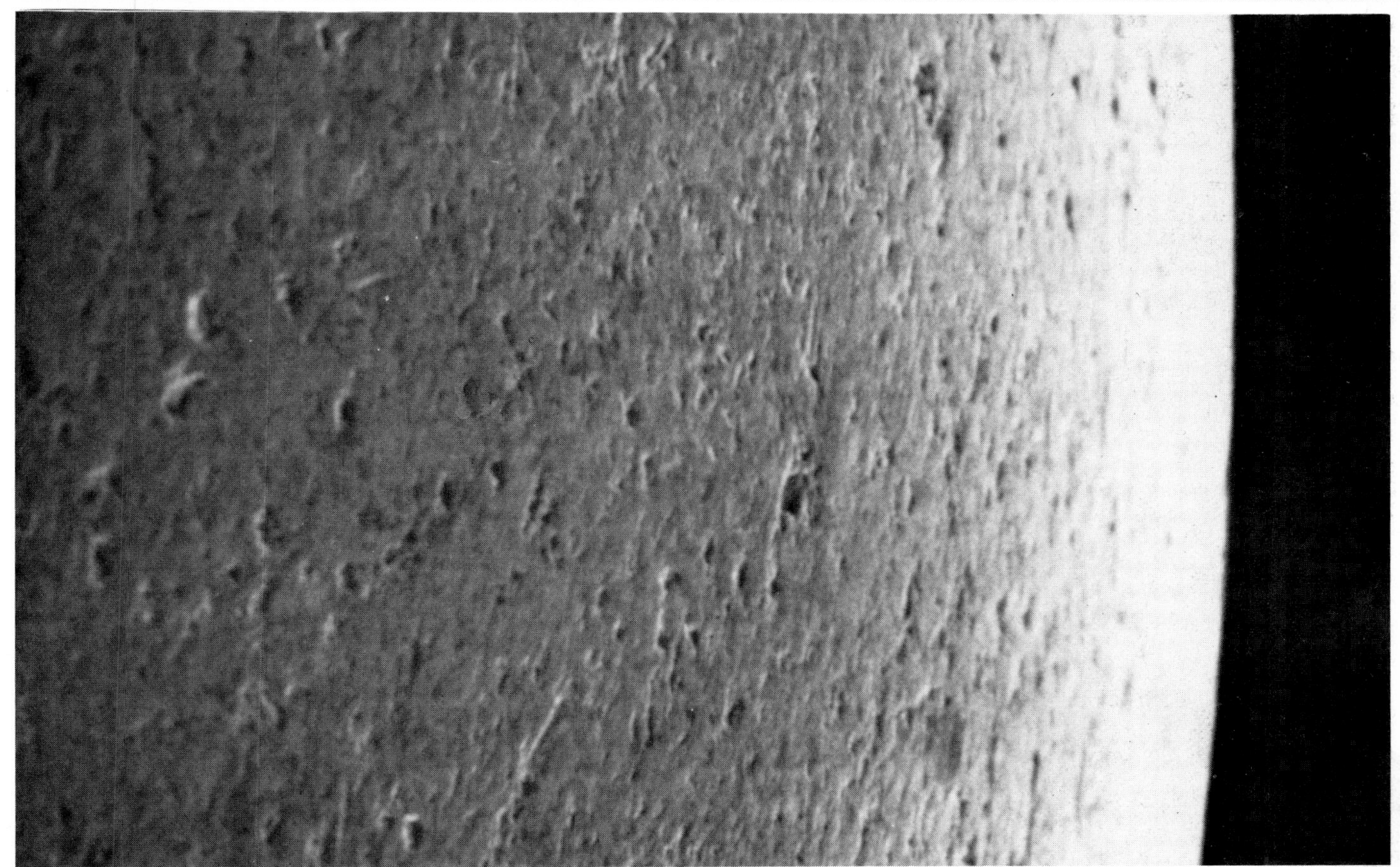

Razor blade, new
4500 X—Eastman Kodak Company, Industrial Laboratory

Razor blade, used
4500 X—Eastman Kodak Company, Industrial Laboratory

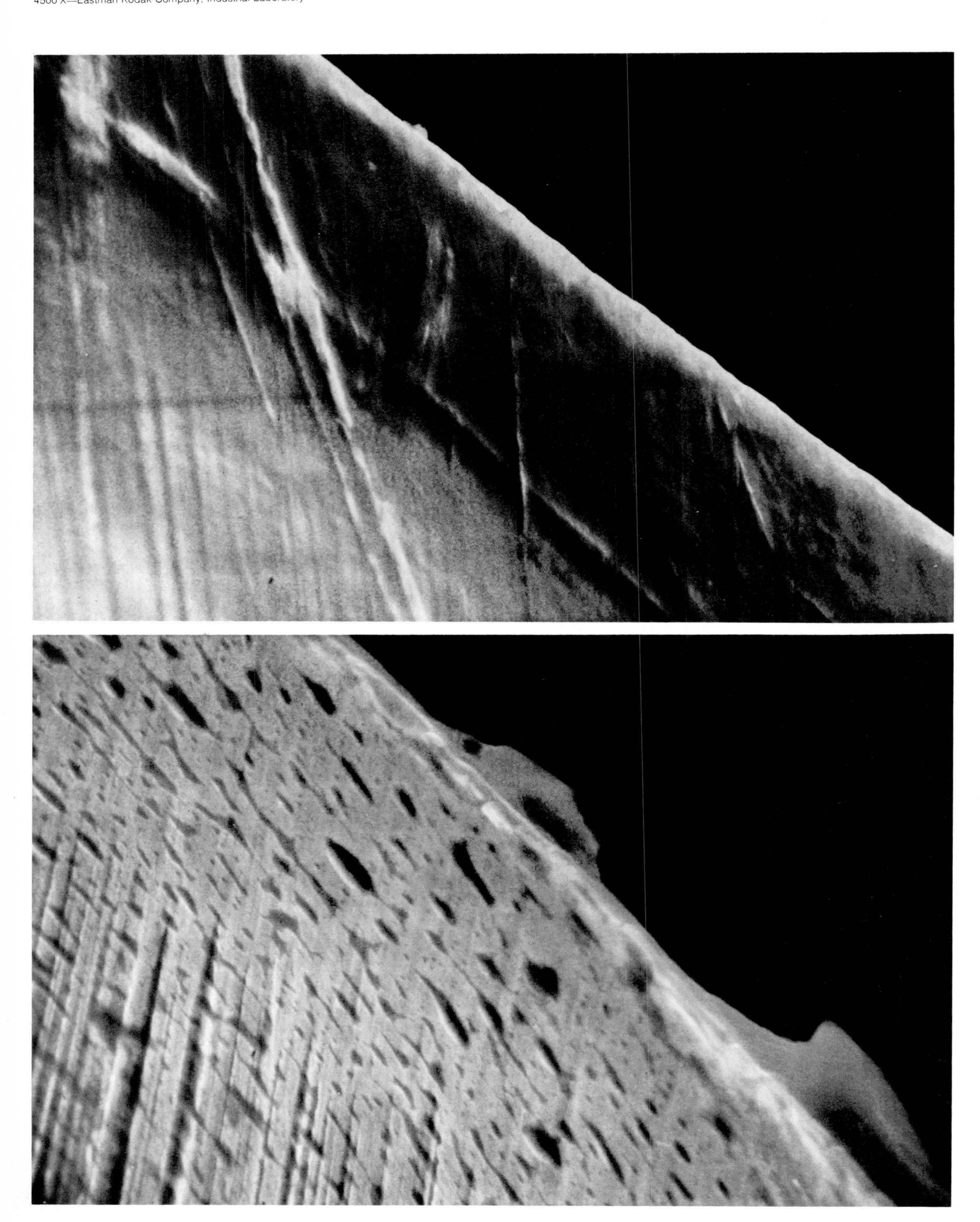

Diamond-studded dental drill
105 X—Eastman Kodak Company, Industrial Laboratory

Phonograph record
230 X—Eastman Kodak Company, Industrial Laboratory

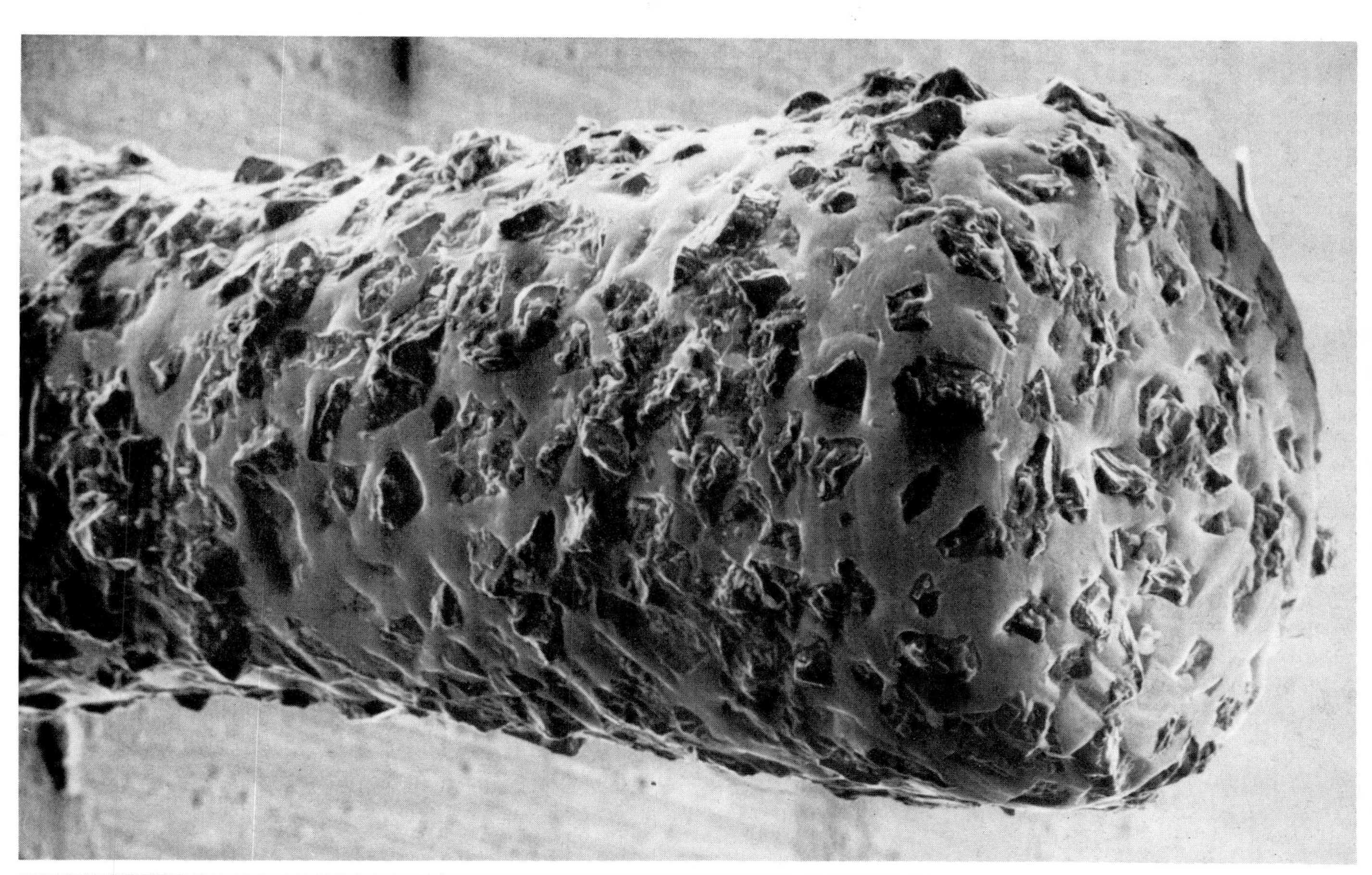

Meat and Potatoes

Muscle meat—in this case stewing beef. The photograph clearly shows the long bundles of muscle fibers.
4200 X—Dr. L. M. Beidler, Florida State University

A single storage cell of a potato. The spherical and egg-shaped objects are starch granules. ►
3000 X—Dr. L. M. Beidler, Florida State University

Paper toweling
600 X—Eastman Kodak Company, Industrial Laboratory

Wasp's nest
800 X—Eastman Kodak Company, Industrial Laboratory

Cotton fabric
220 X—Johnson Wax Photo

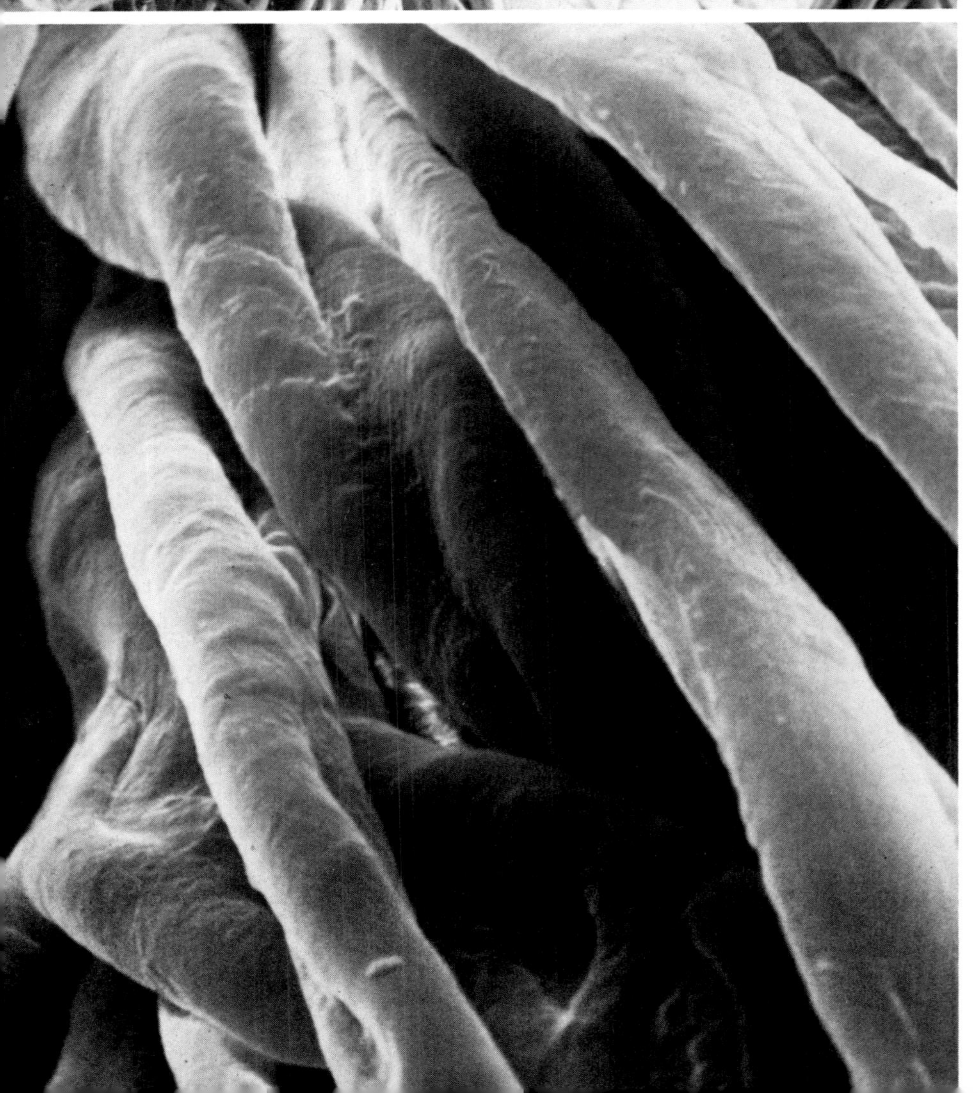

Resin-treated cotton fabric
1550 X—Dr. Emil Bernstein, Eila Kairinen, Gillette Research Institute

Dacron fabric
210 X—Johnson Wax Photo

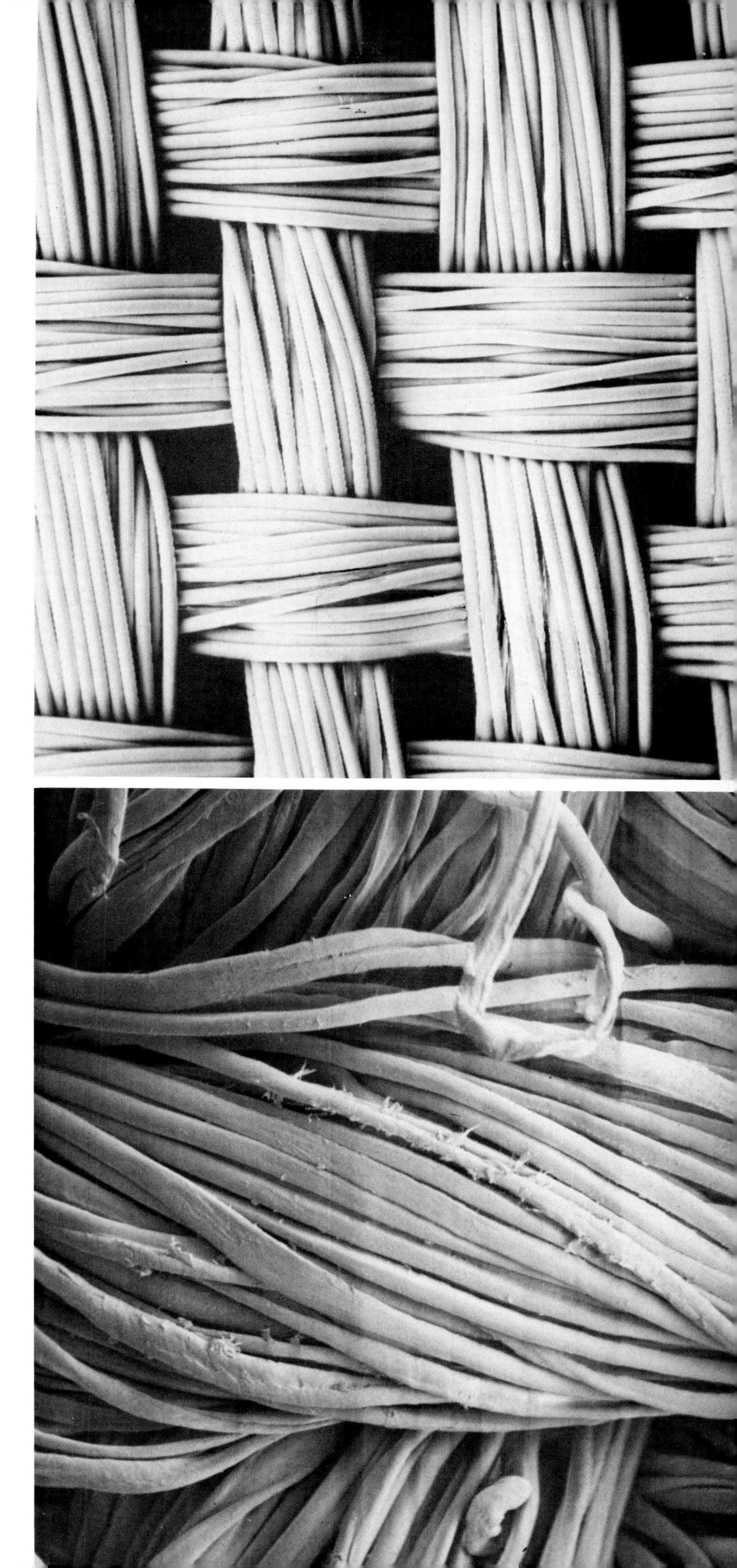

Fabric of 65% Dacron, 35% cotton
475 X—Johnson Wax Photo

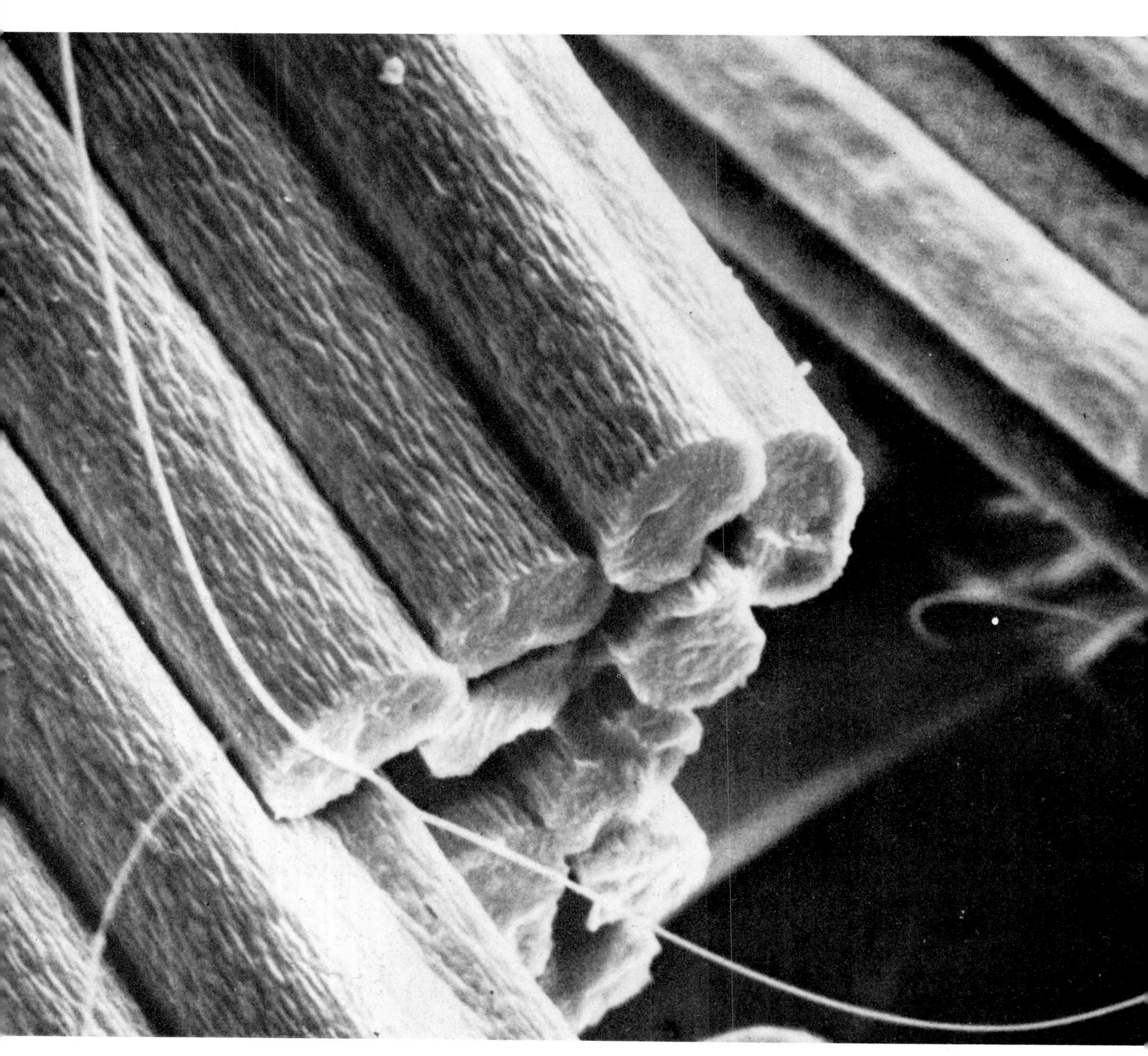

Broken elastic girdle fiber
950 X—Richard Turnage, Advanced Metals Research Corporation

Synthetic hair from wig
1430 X—Courtesy Eastman Kodak Company, Industrial Laboratory

Wool
200 X—Johnson Wax Photo
1000 X—Johnson Wax Photo

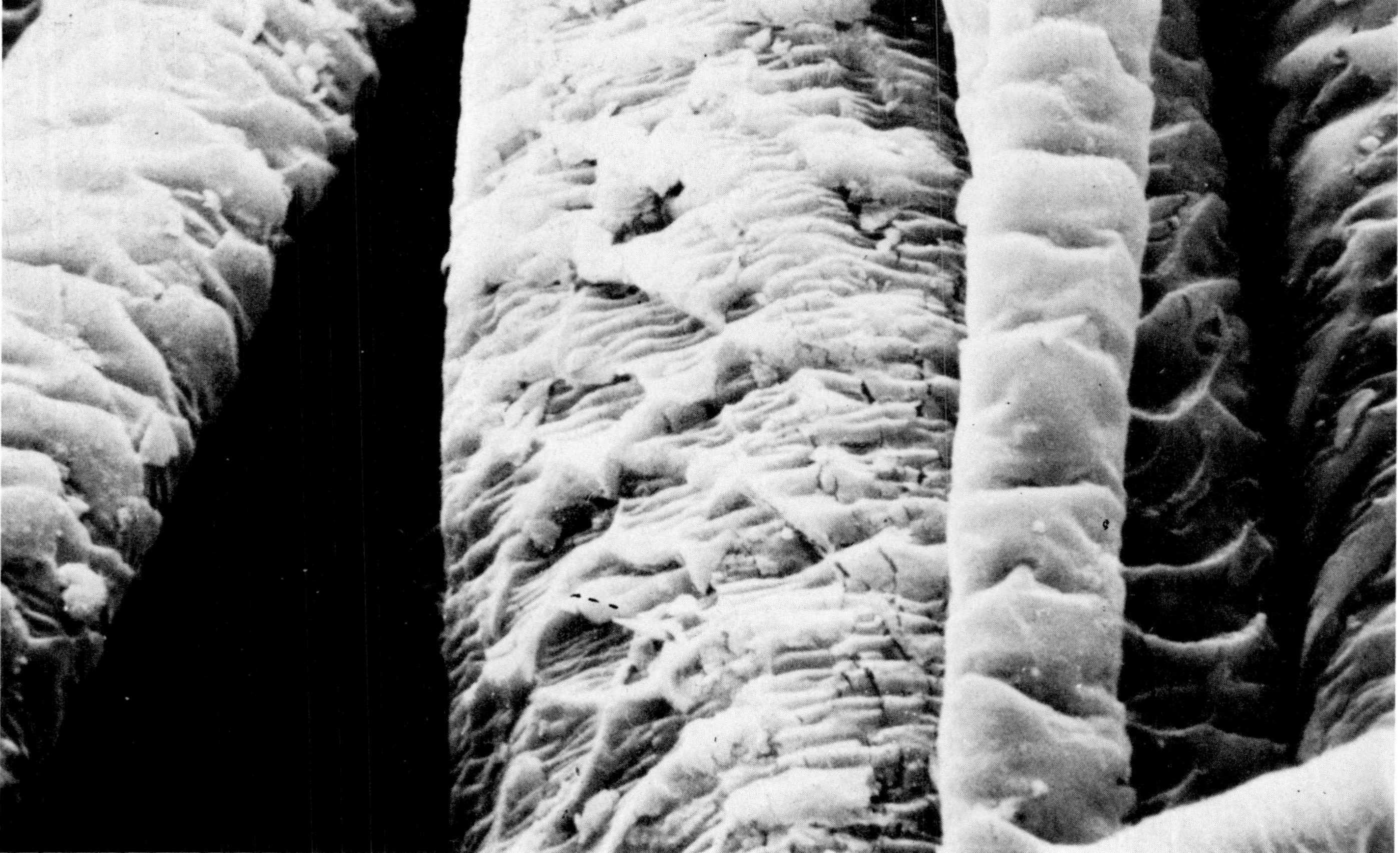

Nylon
400 X—Johnson Wax Photo

Indoor-outdoor polypropylene
425X—Johnson Wax Photo

CHAPTER 4

function follows form

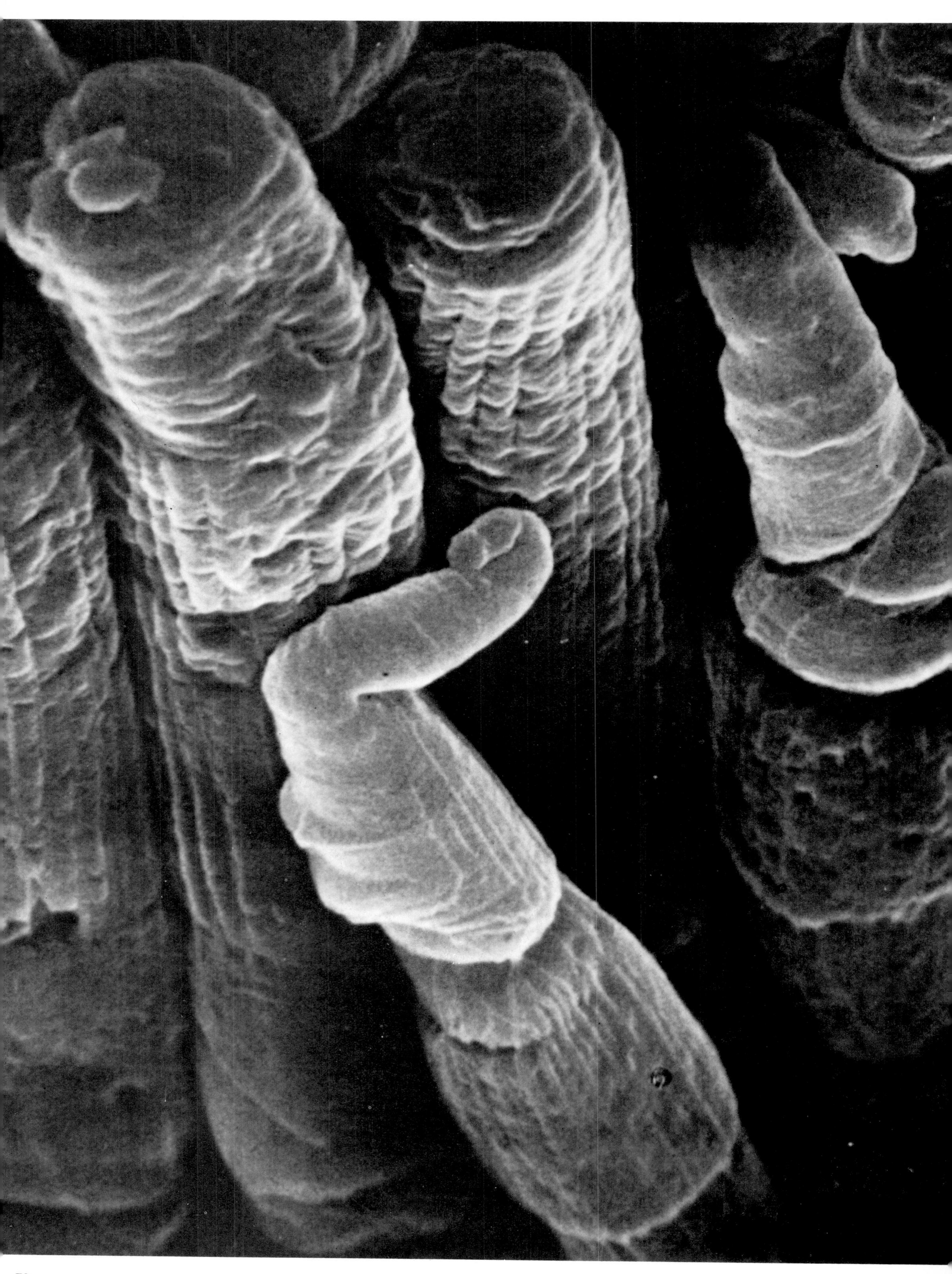

Sensors

Looking down on the retina. The rods, used primarily for dim-light sensing, occupy a much larger part of the total surface exposed to light than do the cones. Thus they can collect enough light energy to respond to faint stimuli. In contrast, the cones (smaller, pointlike objects between the blunt rod ends) need less light collecting area since they operate primarily in bright illumination.
5000 X—Richard Turnage, Advanced Metals Research Corporation

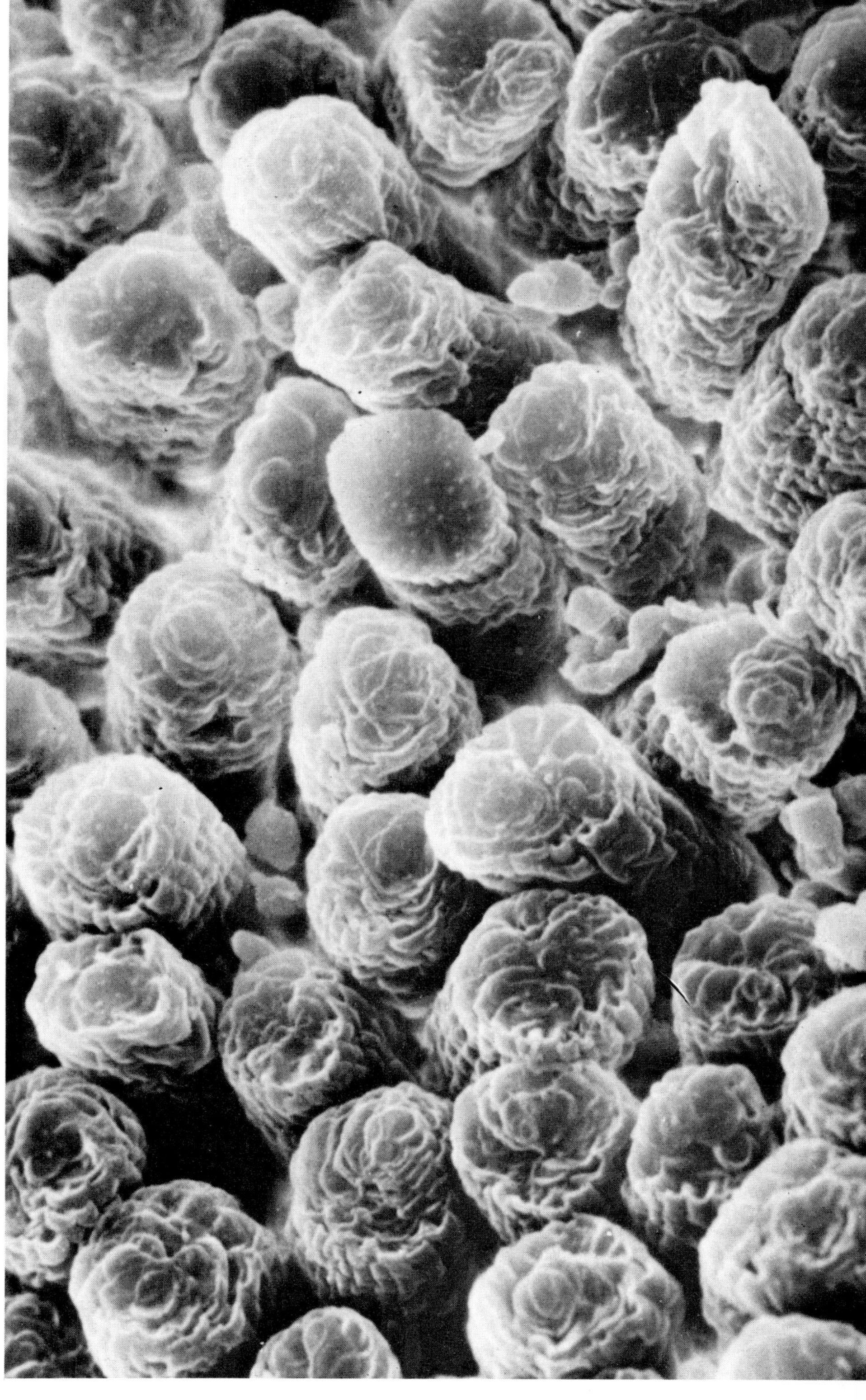

Some of the most remarkable structures found in living creatures are the sensors—the organs by which they see, hear, taste, and otherwise recognize and classify the physical universe. Because sensory input is a steady and never-ending process, we usually take it for granted as a natural phenomenon and think little of the actual mechanisms by which the body gains knowledge of its surroundings. Yet in truth the sensors wonderfully demonstrate the complexity, elegance, and beauty inherent in nature's solutions to the problems of survival. Since the ultimate structures involved in any sensory apparatus are minute, the scanning electron microscope has become a valuable tool in observing their forms in a new and revealing way.

Sight and the eye

Vision is a cooperative venture involving the eye and the brain. It begins when the lens of the eye—the cornea—focuses a picture on the retinal lining at the back of the eye, exactly as a camera lens focuses an image on film. In the retina, photo-receptor cells—the rods and cones—transform changing light intensity and color into electrical signals that are sent to the brain. The human eye functions over an extremely wide range of light intensities—many times wider than the range encompassed by photographic film, for example. It does so by employing two distinct and relatively independent systems: one operating in bright light, the other in dim. The rods, which respond principally to dim light, contain a photosensitive chemical called rhodopsin or visual purple. Its reddish color, easily bleached by light, returns in darkness. This shifting chemical balance caused by light fluctuations generates changes in electrical potential within the rod which are transmitted through a network of nerve cells to the brain. The cones apparently contain a similar but colorless substance. They are far less sensitive than the rods and function primarily in bright light. But unlike the rods, they respond sensitively to different wavelengths or colors of light and account for most color perception. The rods distinguish poorly between colors, which accounts for the fact that objects appear primarily as shades of gray in dim light.

◀ **One of these cones has partially collapsed; another on the right is intact. Rods can be seen in the background.**
8600 X—Dr. Edwin R. Lewis, Dr. Yehoshua Y. Zeevi, Dr. Frank S. Werblin, University of California, Berkeley

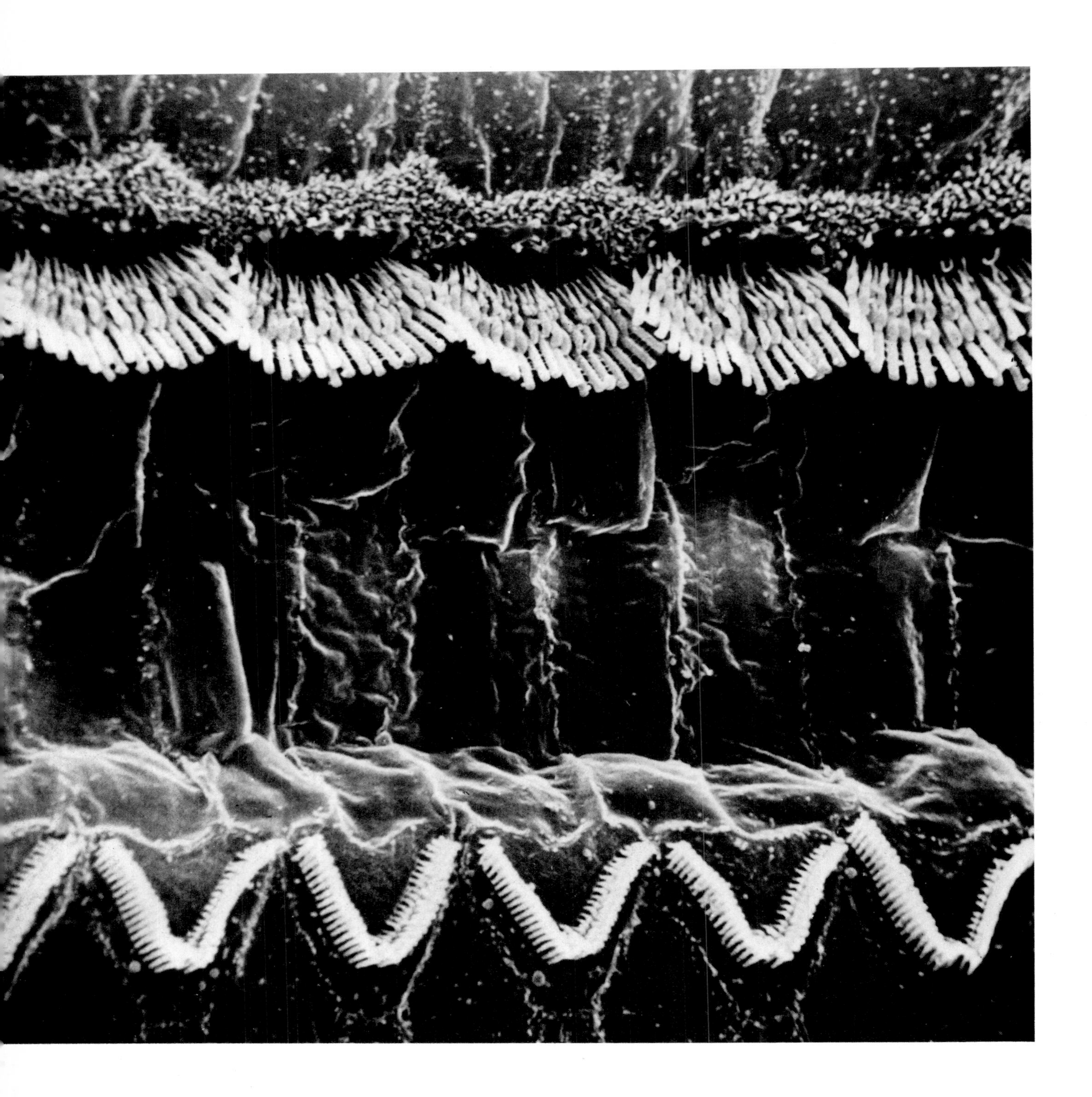

Hearing and the ear

The ear performs two vital functions: it detects the rapid fluctuations in air pressure that we recognize as sound, and it senses motion and direction so that the brain can maintain bodily equilibrium. To perform these functions, it contains two exquisitely designed sets of sensors.

The fan-like clusters at the top are made up of inner hair cells; the V's at the bottom are outer hair cells. They rest on a complicated structure called the organ of Corti deep within the inner ear. The organ of Corti, in turn, is located on the basilar membrane, which runs through the spiral-shaped cochlea. When a sound-pressure wave strikes the eardrum, the motion it causes is transmitted through a network of small bones and a fluid, the perilymph, to the basilar membrane. As the membrane moves, the inner and outer hair cells wave back and forth, suspended in a fluid that tends to remain stationary.

7,800X –Dr. Göran Bredberg, Akademiska Sjukhuset, Uppsala, Sweden

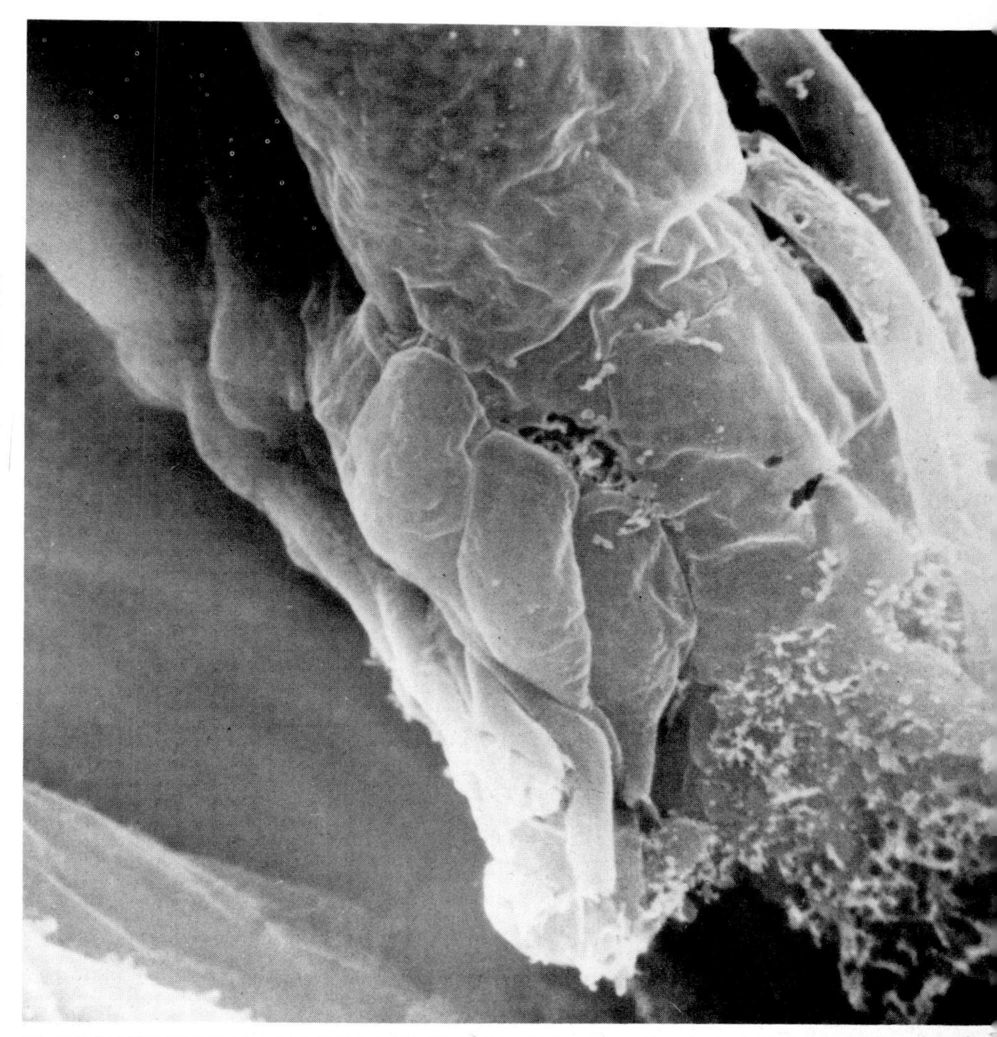

As each hair cell moves, it generates small electrical signals which are transmitted to the brain and there interpreted as sound. The large cylinder disappearing at the top of the picture is an outer hair cell. The several flattened structures attached to the base of the hair cell in the center of the photo are nerve endings that sense the changing electrical potential in the sensor cell.

11,200X Dr. Goran Bredberg, Akademiska Sjukhuset, Uppsala, Sweden

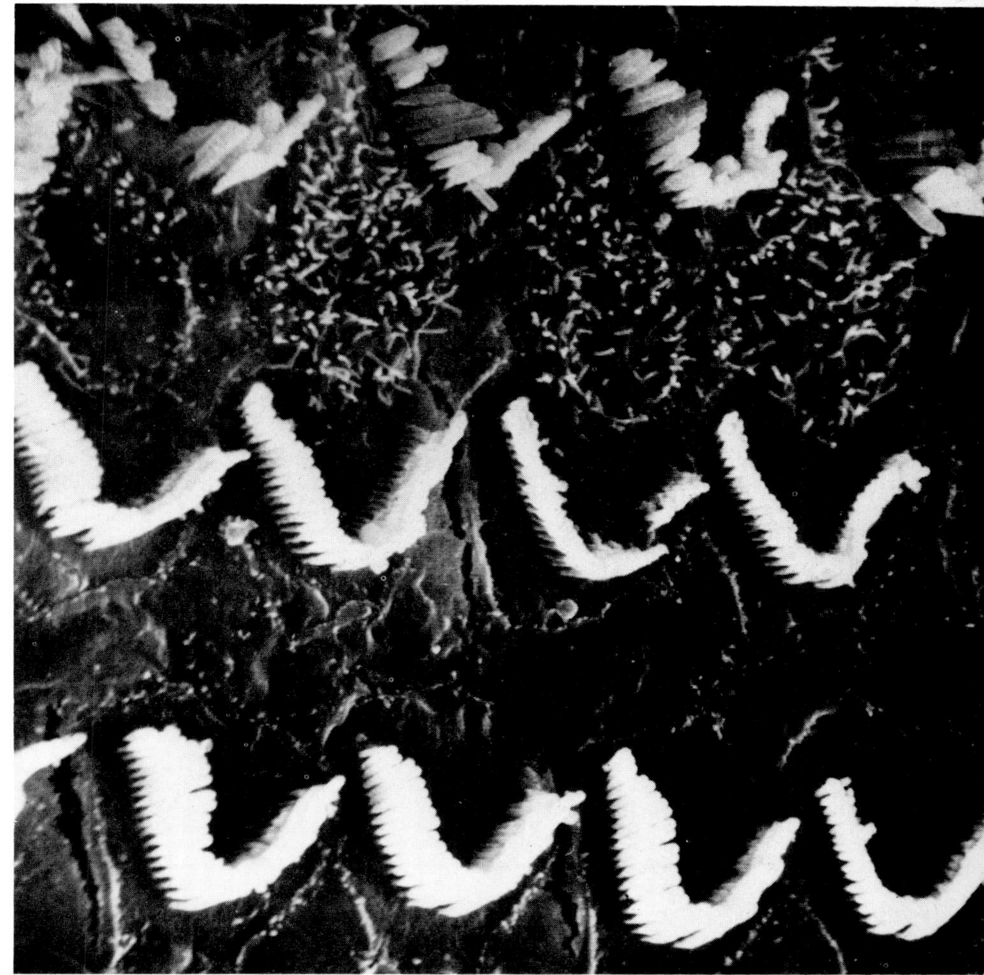

A guinea pig was exposed to high-intensity sound; this is the result. The first row of outer hair cells is severely damaged, the center row somewhat less so.

3600 X—Dr. Goran Bredberg, Akademiska Sjukhuset, Uppsala, Sweden

These are motion sensors from a region of the ear of a bullfrog (*Rana catesbeiana*) known as the sacculus. Near the top of each clump of cilia can be seen a small bulbous object. This structure connects the cilium directly beneath it—called the kinocilium—to the many similar but distinct stereocilia that make up the remainder of the cluster. Except for the bulb at the end, the kinocilium looks precisely like all of the other cilia. But it differs in several important ways, and is the key element in the animals's motion-detection system
4200 X—Dr. D. E. Hillman, Dr. Edwin R. Lewis, University of California, Berkeley

The ear serves not only for hearing, but also for balance. To perform this function, it contains sensors that detect changes from a resting to moving state or changes in the rate of speed (acceleration). In principle, these detectors are not unlike the accelerometers that perform the same function in airplanes and space vehicles. But in construction and operation, they are totally different. A spaceship's motion sensor is a gyroscope whose spinning rotor remains pointed in the same direction no matter how the vehicle around it moves. The ear achieves the same end with these minute bundles of hairlike cilia. This striking series of photographs has finally revealed how they do it.

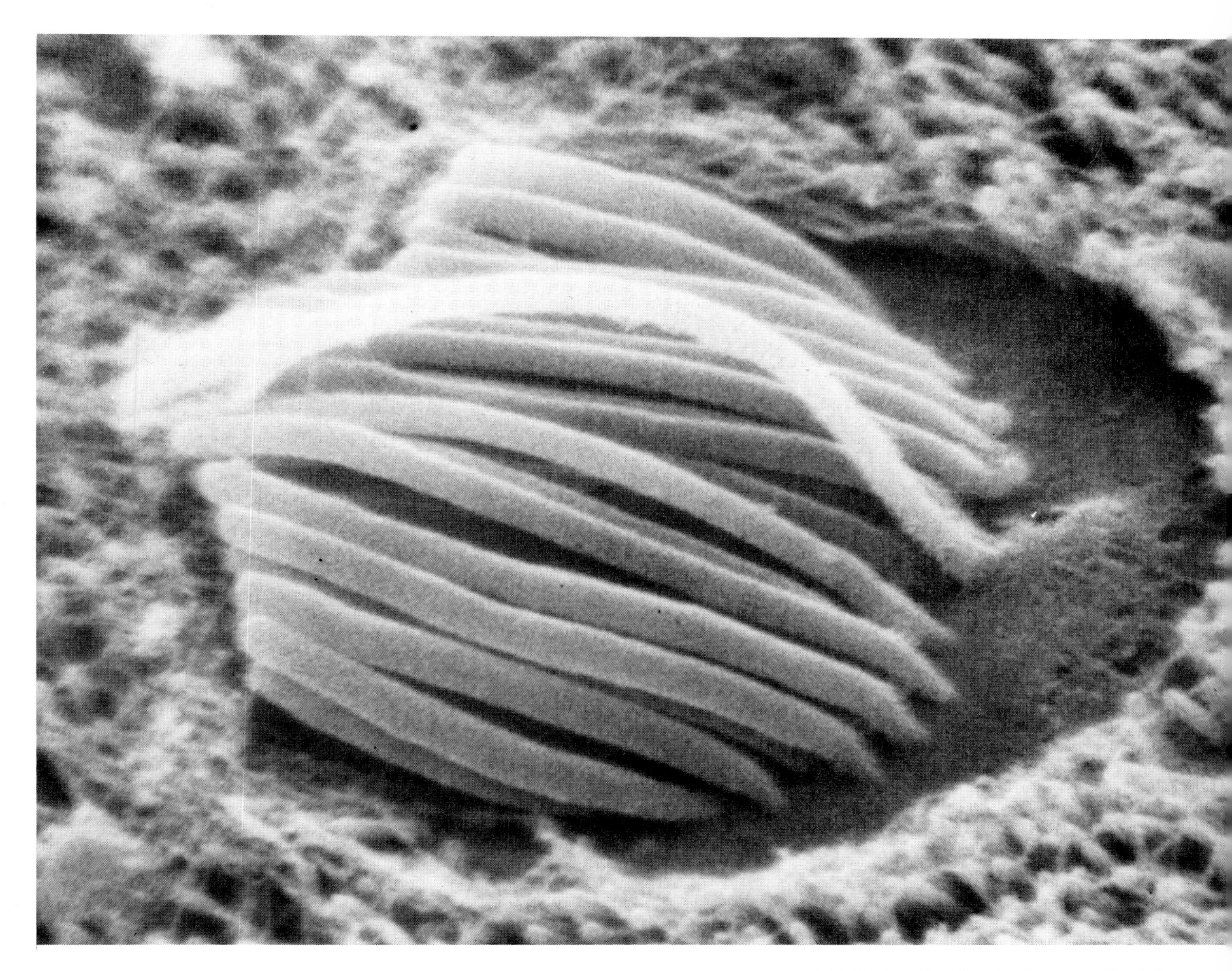

This closeup of a ciliary bundle shows the distinctive character of the kinocilium more clearly. (This specimen is from the sacculus of a mudpuppy, or salamander. It has no prominent bulb at the top, but it is identical to the frog's sensing mechanism in all other important details.) As can be seen, the base of the kinocilium is rooted in its own area, slightly apart from those of the stereocilia. The stereocilia are implanted in a relatively hard, rigid cuticle. But the firm base does not extend under the kinocilium. When an animal moves, the fluid which fills the chambers and in which these cilia grow tends to resist movement because of inertia. Thus between the ear and its fluid there is relative motion which causes these ciliary bundles to flex. As they move back and forth, the base of the stereocilia remains rigid. But the kinocilium, which is attached to the stereocilia at the top, is forced up and down in a plungerlike motion. This stimulates the cell to which the kinocilium is attached, and this cell in turn stimulates nerves connected to it, which send signals to the brain. Ciliary sensors of this type are oriented in various directions throughout the chambers of the inner ear and detect not only motion, but also its direction.

52,200X—Dr. Edwin R. Lewis, Mrs. Paula K. Nemanic, University of California, Berkeley

Taste and the tongue

Despite the infinite variety of flavors which seem to be identifiable, the taste buds of the human tongue are actually able to detect only four basic tastes: sweet, salty, sour, and bitter. The distinctive flavors of individual foods develop out of the proportions in which these four elements are combined, along with the far more numerous and subtle aromas that are detected by the nose and combined by the brain into the overall impression of flavor. The sensation of taste is detected within the taste buds, each of which contains a small pore through which samples of food enter. Although the process is imperfectly understood, ions and molecules of the sample are believed to react electrically and chemically with taste cells. These reactions are sensed by nerve cells, various kinds of which react to certain tastes or combinations of tastes. Hydrogen ions are thought to react to sour tastes; ions of chlorine, bromine, and iodine to saline tastes; molecules in the alkaloids of quinine and other substances to bitter tastes; and alcohols and sugars to sweet tastes.

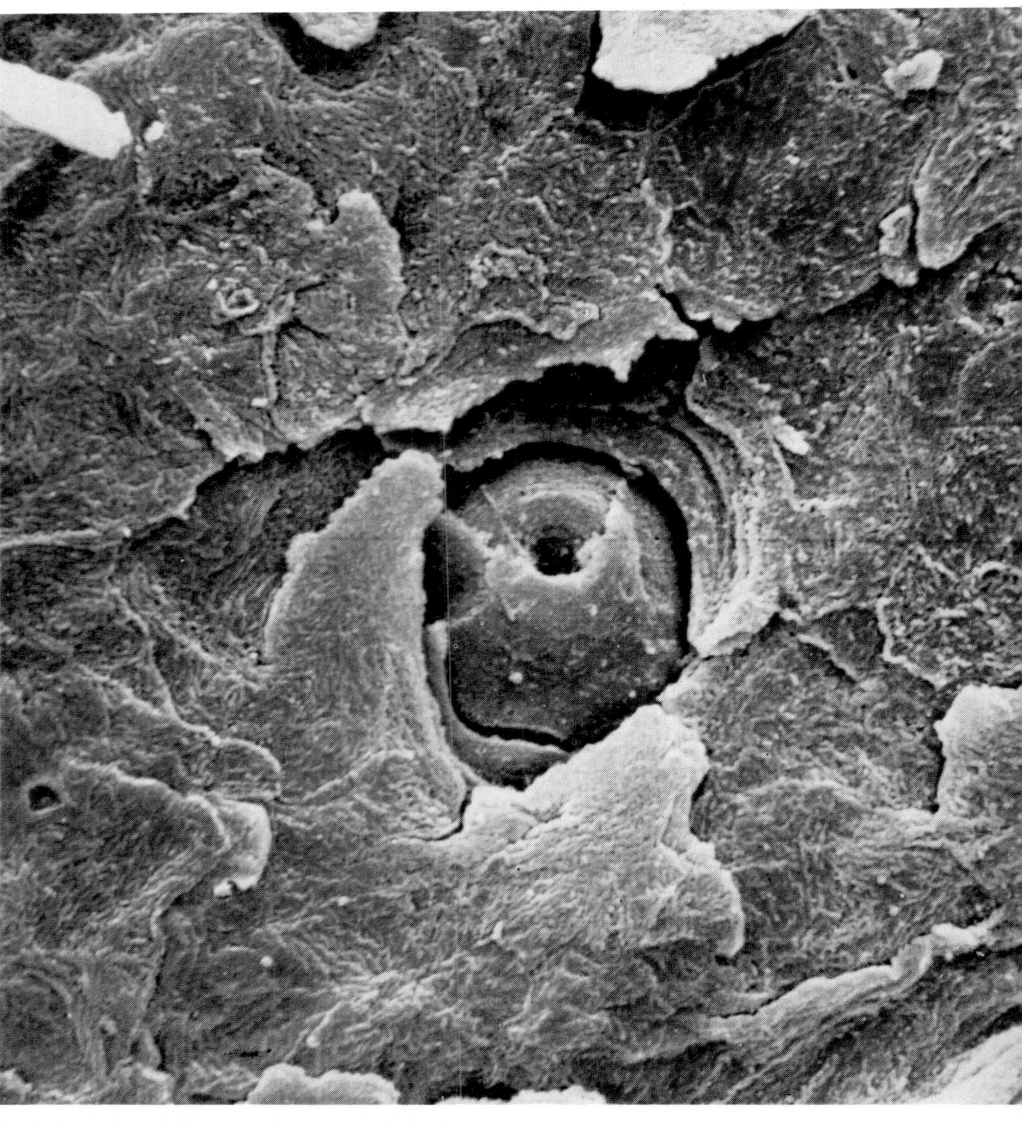

Taste bud of a rat. The pore through which food enters is clearly visible at the center. Each of the sheetlike elements covering the surface of the tongue is a single cell.
2,600X–Dr. L. M. Beidler, Florida State University

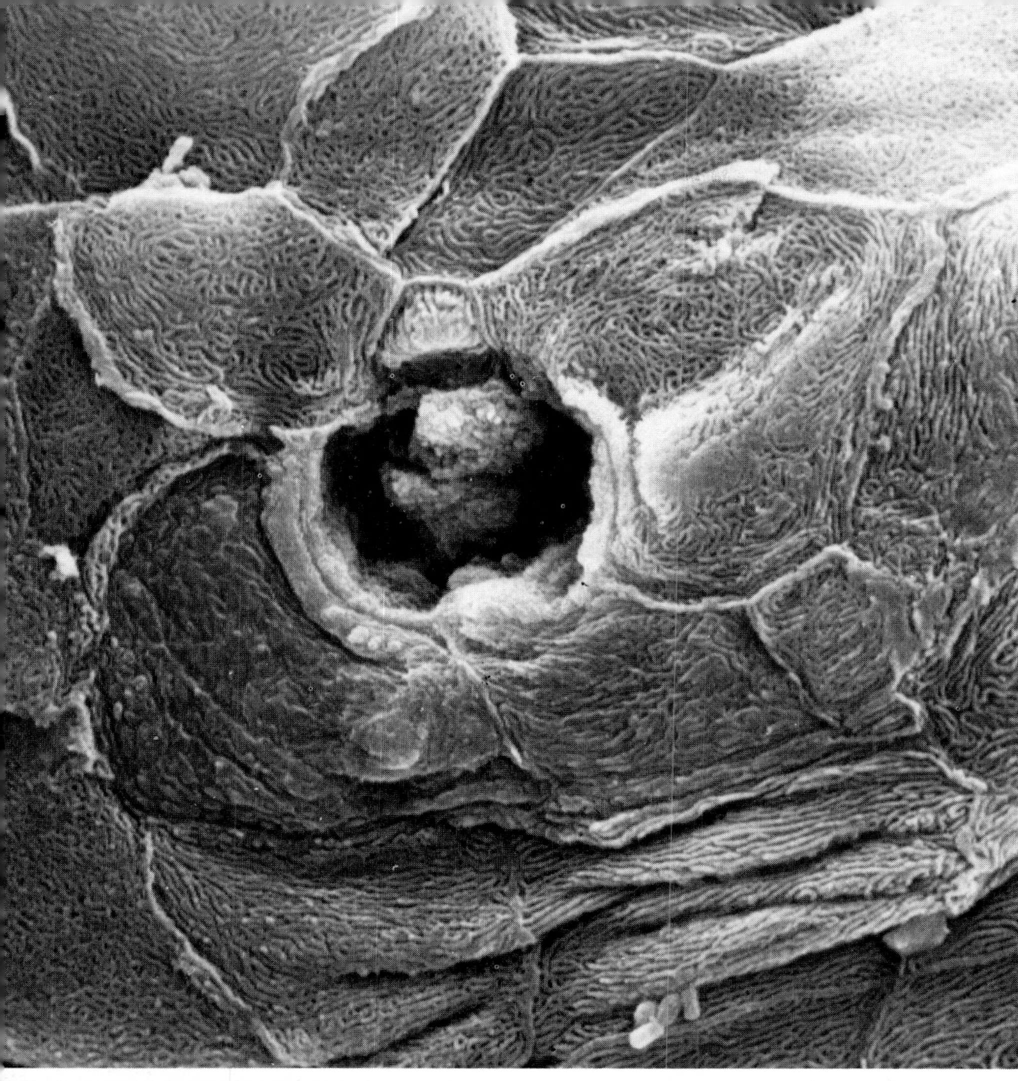

Human taste bud. The general structure is strikingly different from that of the taste buds in the previous photographs; the tongue is covered not by protruding papillae, but by soft, overlapping sheetlike cells, each with its own intricate surface pattern which is doubtless as distinctive as a fingerprint.
2,800X –Dr. I. Kaufman Arenberg, Washington University School of Medicine

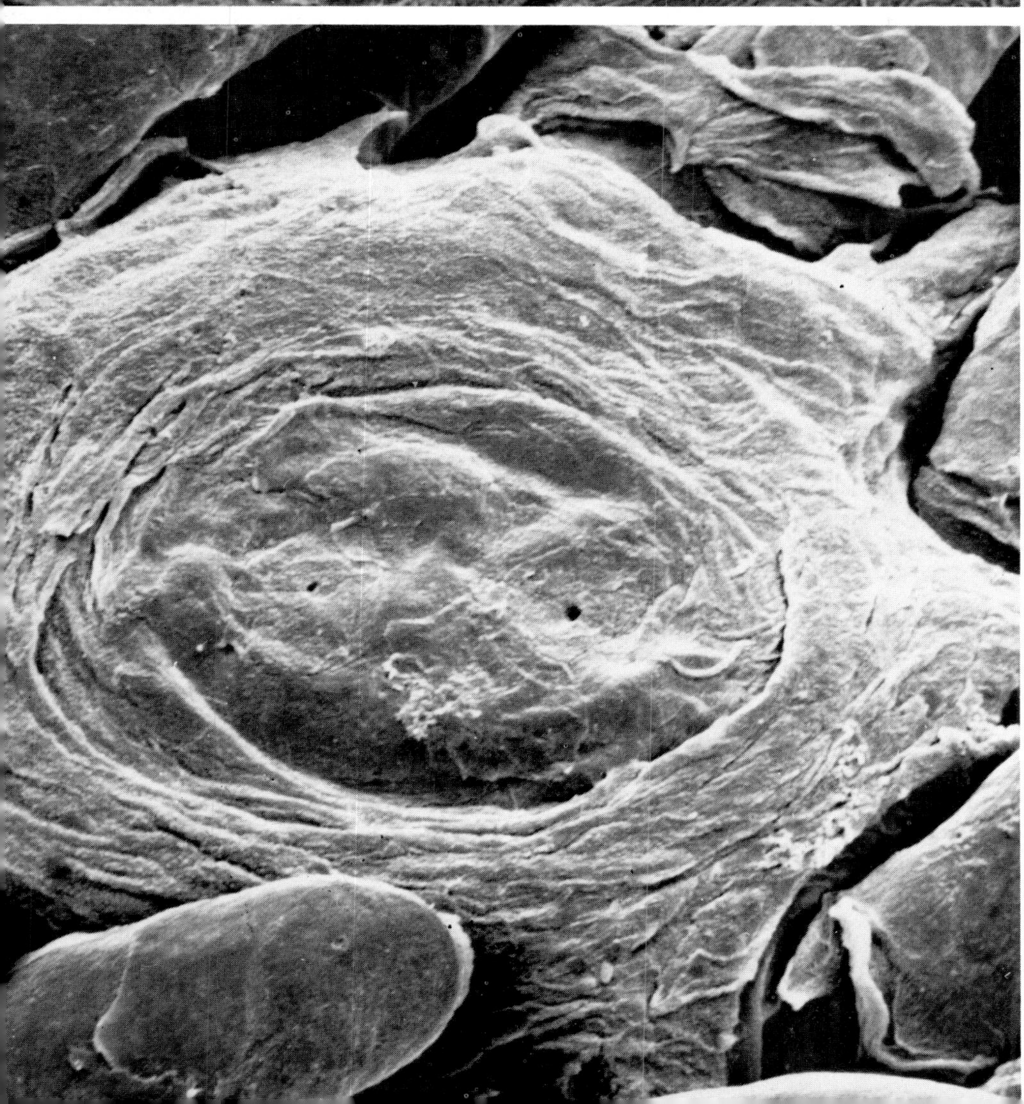

Two fine pores that lead to the taste cells can be seen in this taste bud on the tongue of a vampire bat.
1,300X –Dr. Irving Fishman, Grinnell College

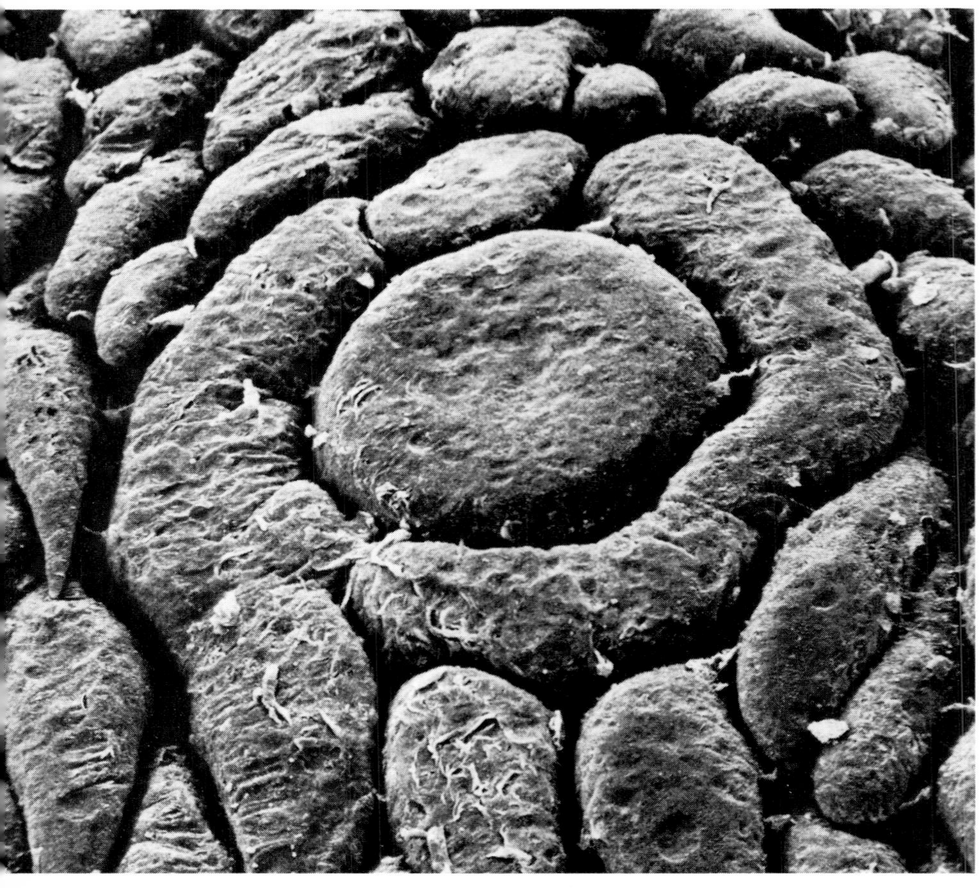

The tongue of a three-week-old puppy contains flat, soft elements called circumvallate papillae. Taste cells are in the trenches.
115X –Dr. L. M. Beidler, Florida State University

A guinea pig's tongue is raspy, rather than smooth like the puppy's. This picture shows why.
180 X—Dr. I. Kaufman Arenberg, Washington University School of Medicine

Unkindest tongue of all: the shark's. These denticles—like small teeth—can abrade like sandpaper. The denticle at center is just emerging to replace a lost one in this space.

195X —Dr. Sarah A. Luse, College of Physicians and Surgeons, Columbia University

Chromosomes

Every individual cell of a human body contains the complete blueprint for the entire being—from the color of the eyes to the exact shape of a big toe. This enormous amount of information is contained in the nucleus of the cell in a set of structures called chromosomes. When a cell divides, each chromosome within it splits into two chromosomes, so that each of the resultant cells is a perfect replica of the original. Thus do life and growth take place. This scanning electron microscope photograph shows chromosomes from a Chinese hamster in the process of division.
4,200X –Dr. K. M. Marimuthu, New England Medical Center Hospitals

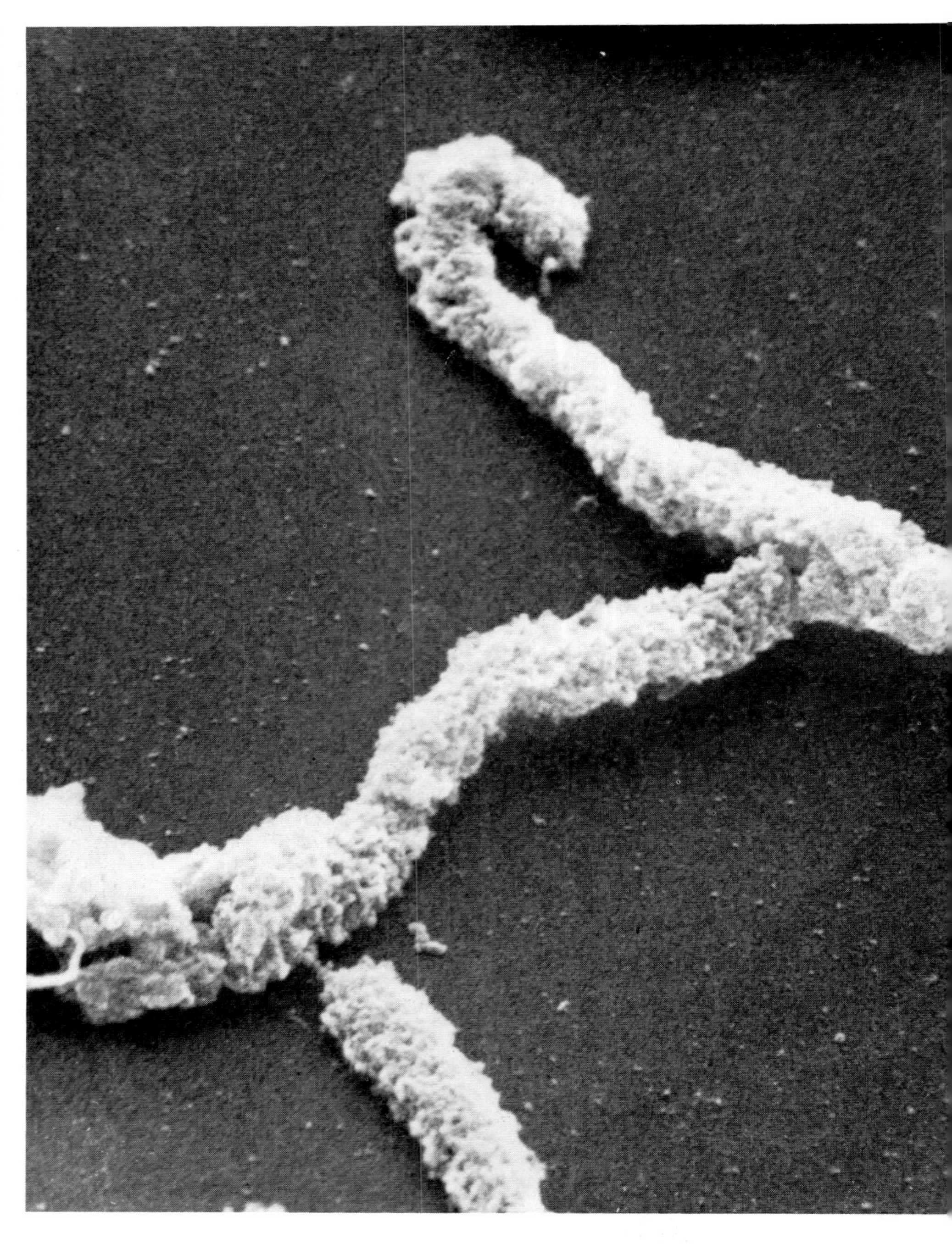

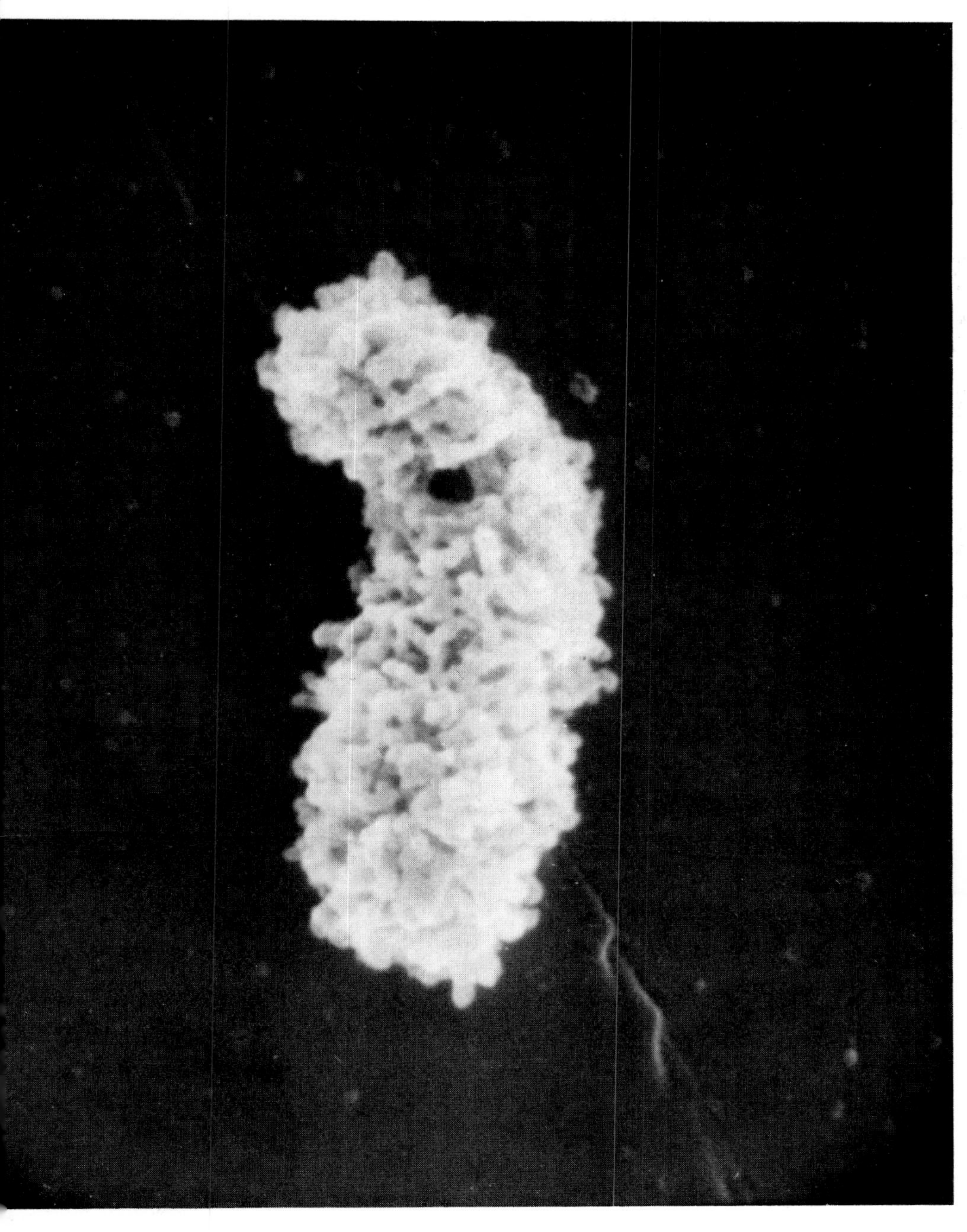

A closeup of a chromosome shows the chromosome fibers or chromomeres which contain the actual genes or units of heredity.

13,000 X—Dr. K. M. Marimuthu, New England Medical Center Hospitals

Cilia

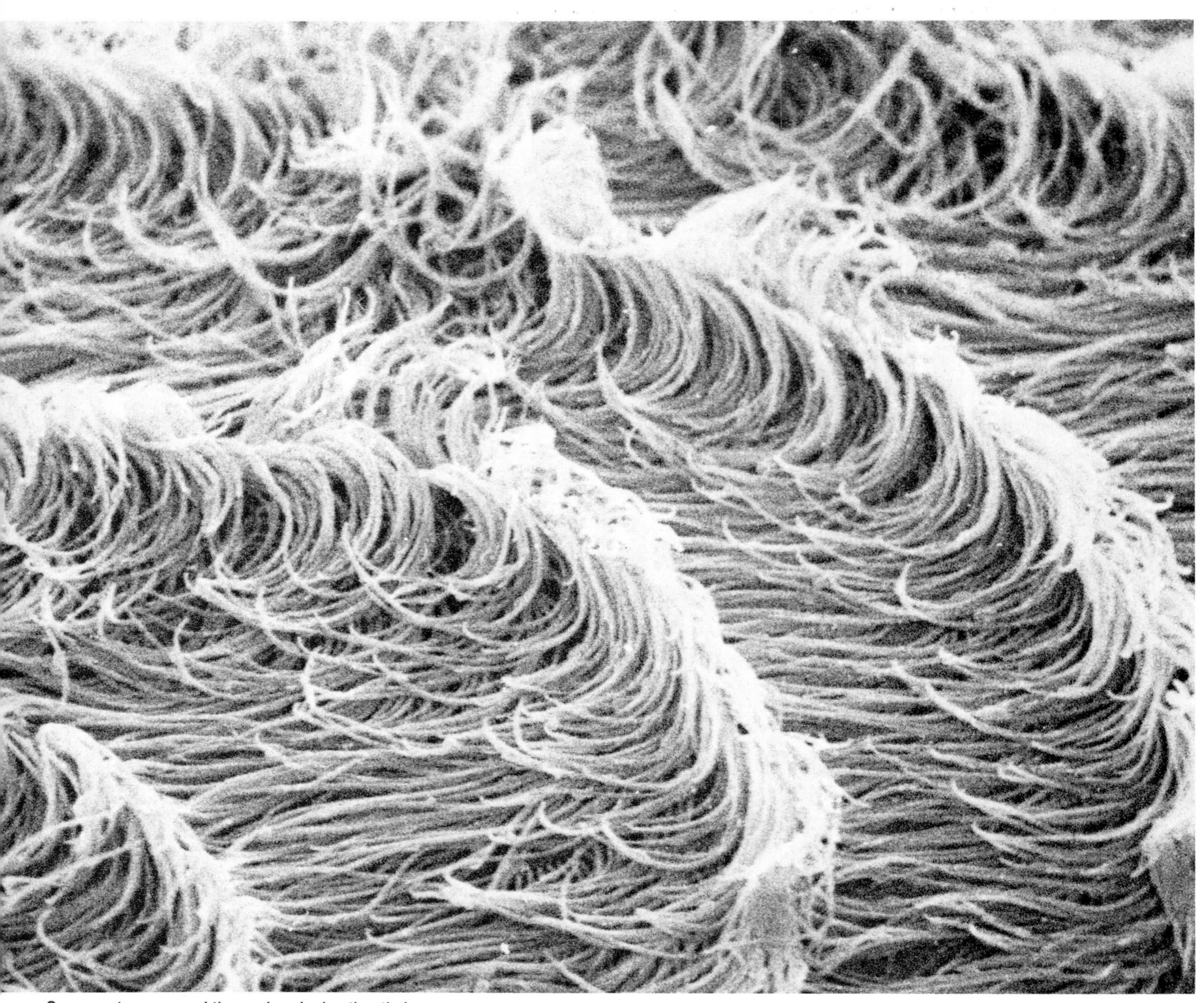

Some protozoa propel themselves by beating their cilia—fine, hairlike strands which cover their bodies—back and forth in a rhythmic, highly coordinated, wavelike motion. Similar cilia within the human bronchial passages beat in much the same way to expel dirt, dust, pollen, and disease organisms from these vulnerable areas. Thus they play a large part in preventing respiratory infection. This remarkable photograph catches the cilia in midbeat. It was made by isolating several hundred *Opalina ranarum*, a ciliated protozoan, in a few drops of fluid in a watch glass, then freezing their motion by flushing them with a fixing solution. Finally, they were washed, dehydrated and dried, capturing the actual wave motion in midcycle. Investigators had known that *Opalina* can change its direction while swimming; this photograph made the mechanism clear. The creature is caught in the process of altering course; the cilia at left (top) are moving in one direction, those at right (bottom) in another. The photograph also revealed to investigators that ciliary motion is not a simple back and forth movement. Each cilium beats forward, rotates counterclockwise and positions itself parallel to the body during its backstroke or recover phase, then straightens up in preparation for the next forward stroke.

3,200X –G. A. Horridge, Australian National University; S. L. Tamm, Indiana University

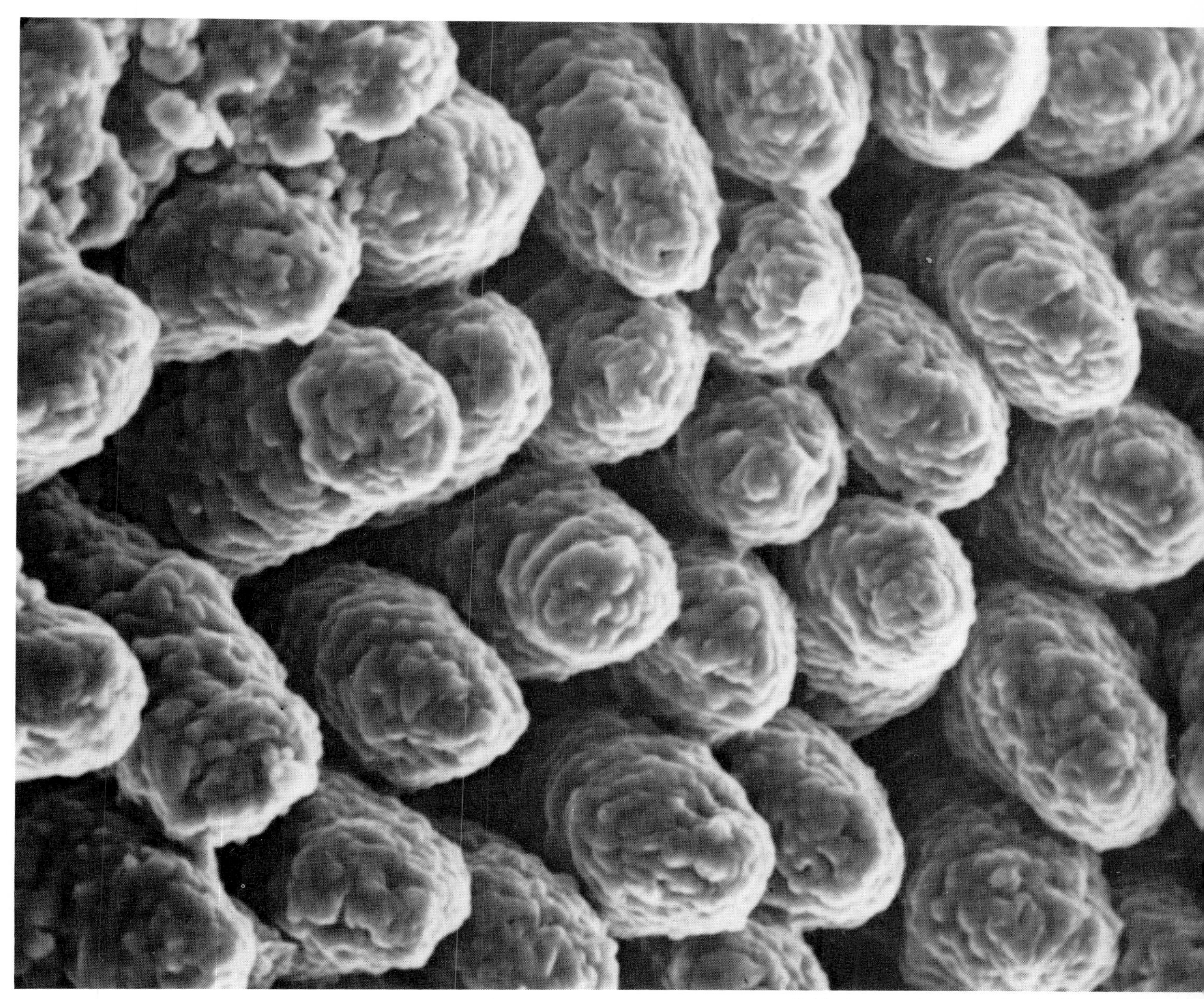

Nutrients, as we know, are absorbed by the body in the intestines. The actual absorption takes place over the tremendous surface area provided by bumpy protrusions that line the intestine, the villae. Carbohydrates and proteins in the form of amino acids are absorbed directly into the bloodstream through the walls of small capillaries that line the villae. Fats are absorbed by a more circuitous route: They are changed to fatty acids and glycerol in the intestine, absorbed by small structures on the villae called lacteals, and ultimately make their way back to the bloodstream.
290X —Miss B. Farber, Johnson and Johnson Research

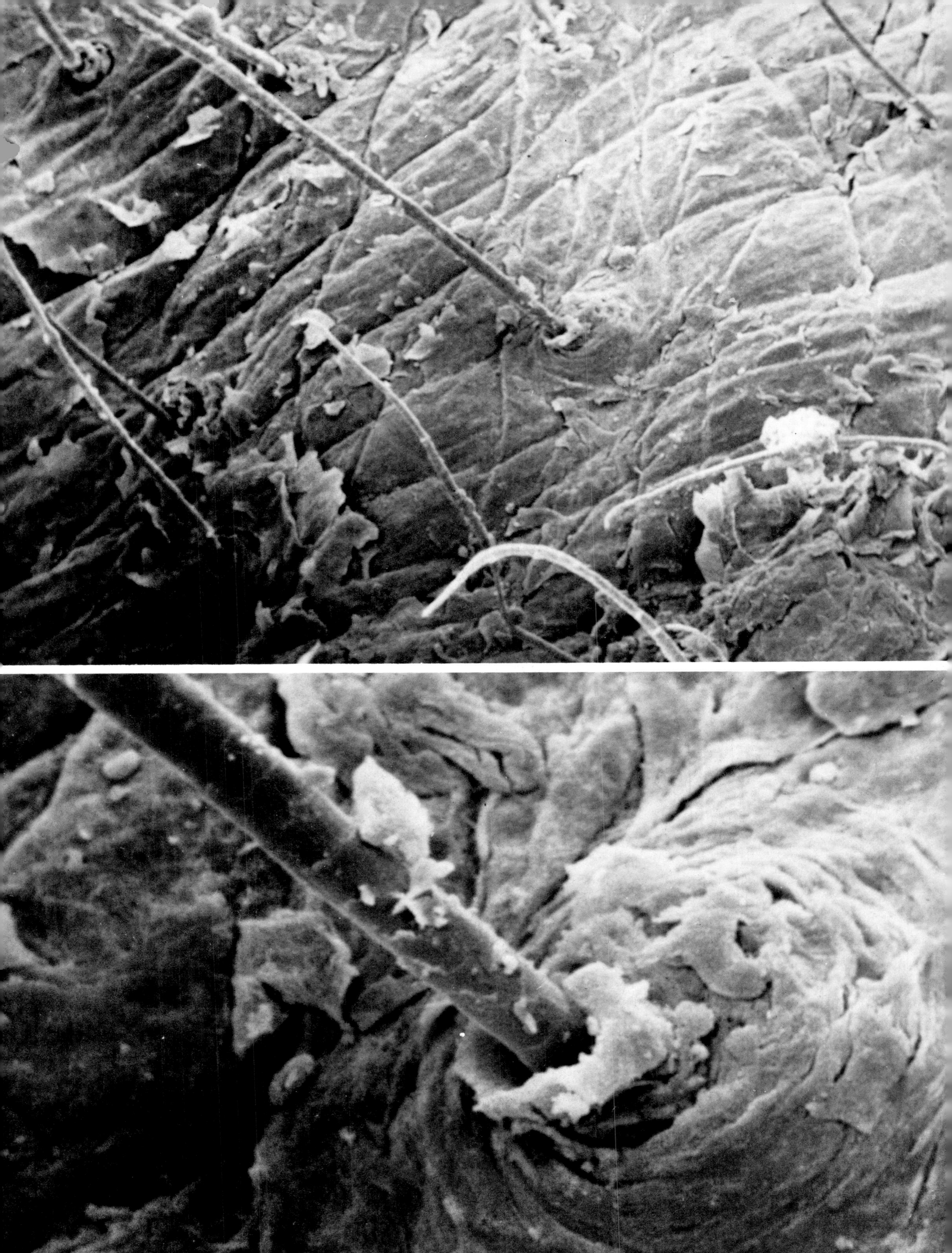

◄ **Human facial skin and hair. As skin cells age, they peel off and are replaced by lower layers. This process of desquamation can be clearly seen.**
210 X—Dr. Joseph Gennaro, New York University

Skin and Hair

Nothing is more familiar than human skin and the various kinds of hair that sprout from it. Yet the scanning electron microscope gives a new view of this well-known terrain.

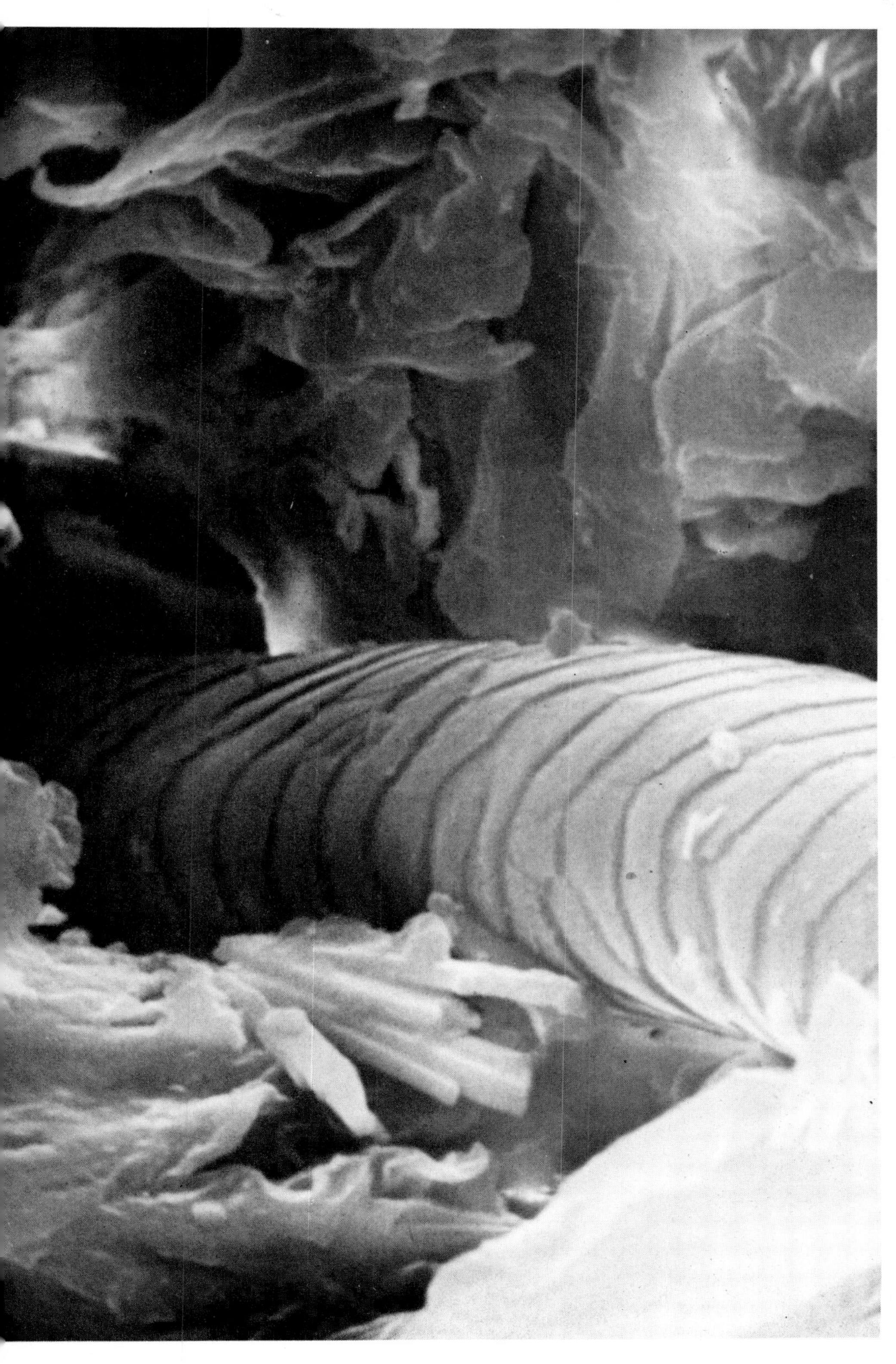

Ultra-closeup of a normal hair growing from the stratum corneum—the outer layer of the epidermis. Note the layered cuticle or outer layer of the hair, also the debris—mostly dead cells—lodged in the follicle opening.
1600 X—Dr. C. M. Papa, Miss B. Farber, Johnson and Johnson Research

◄ **A closer look at a single hair. Just below the surface of the skin, within the hair follicle, are located sebaceous glands. They generate a fatty material called sebum, which oozes up the shaft to lubricate the hair and surrounding skin. A clump of extruded sebum can be seen around the base of the hair.**
1000 X—Dr. Joseph Gennaro, New York University

Replica of a sweaty palm—showing both the ridges that make up the palmprint and the droplets of perspiration being secreted.
75 X—Dr. Emil Bernstein, Eila Kairinen, Gillette Research Institute

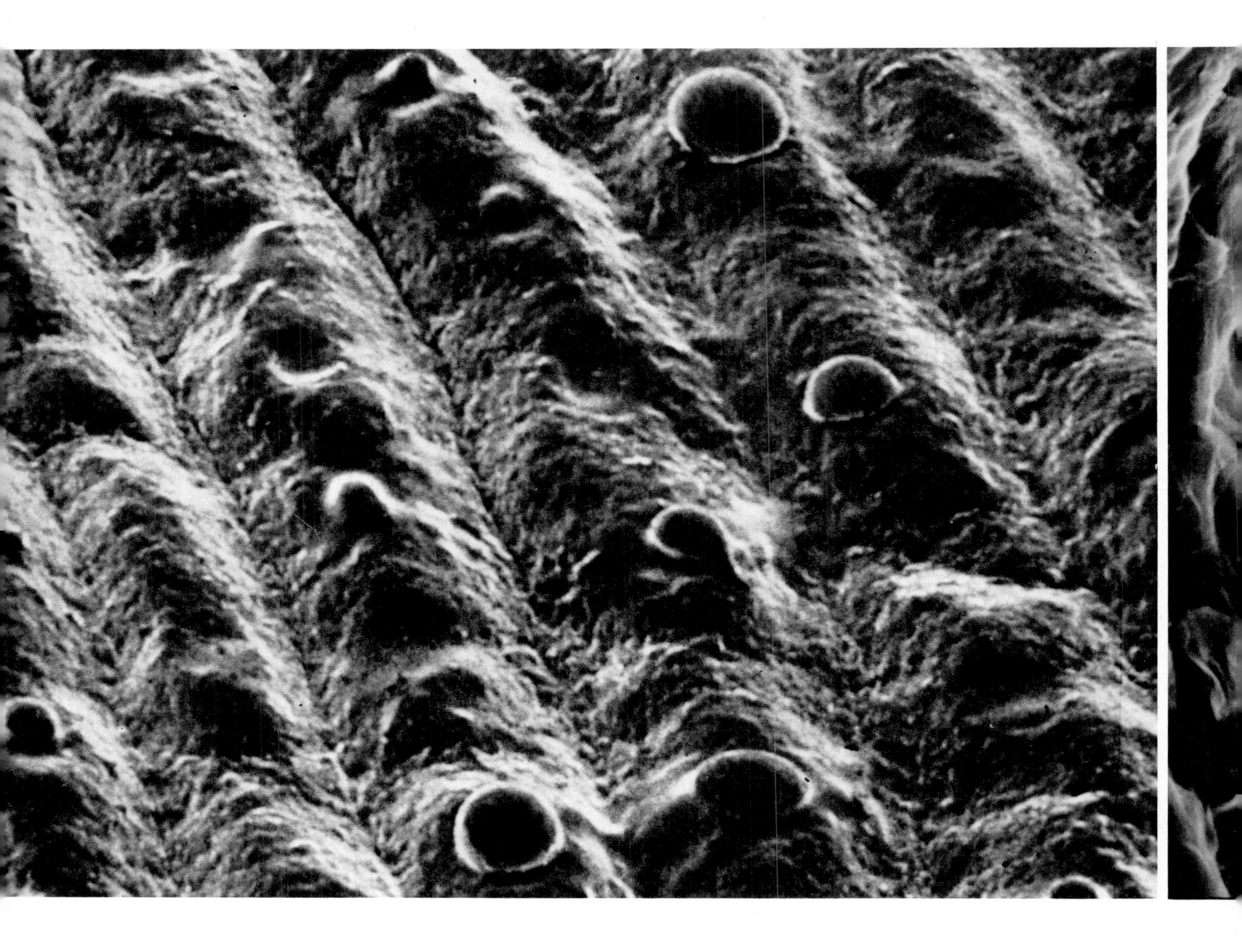

Normal sweat pore
320 X—Dr. C. M. Papa, Miss B. Farber, Johnson and Johnson Research

Psoriasis—pores such as this one clogged with cell debris accompany the red welts and intense itching of this annoying disease.
340 X—Dr. C. M. Papa, Miss B. Farber, Johnson and Johnson Research

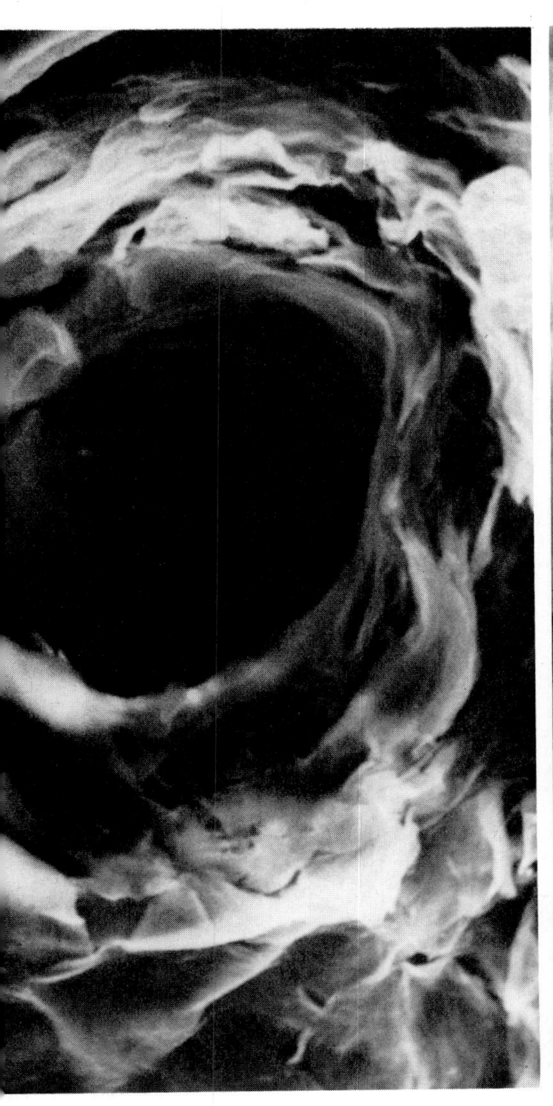

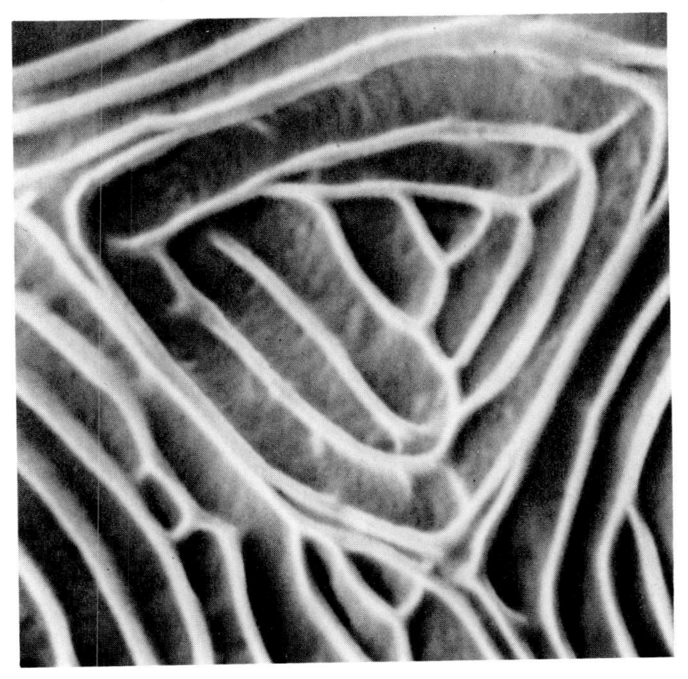

Guppy skin. The double ridges show where one cell abuts another.
12,000 X—Dr. Sarah A. Luse, College of Physicians and Surgeons, Columbia University

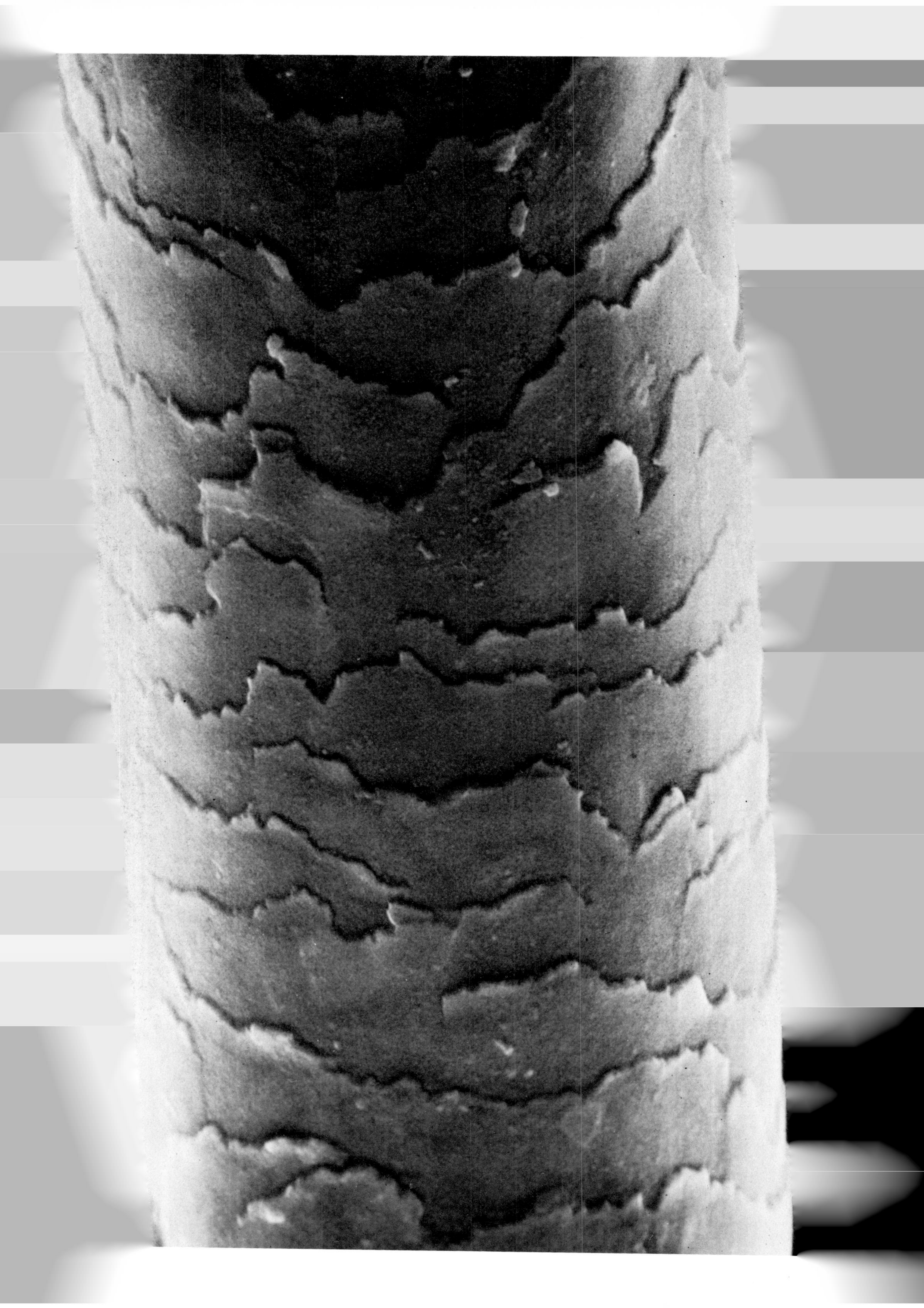

◀ **Extreme closeup of a normal human hair, showing the barklike outer cuticle.**
3300 X—Eastman Kodak Company, Industrial Laboratory

Bleached hair
930 X—Miss B. Farber, Johnson and Johnson Research

Hairs sliced off with a sharp razor blade. The central core—the medulla—the cortex, and the thin outer cuticle layers can be seen. The streaks across the cut faces were caused by imperfections in the blade's cutting edge.
1100 X—Dr. Emil Bernstein, Eila Kairinen, Gillette Research Institute

A company that makes a shaving product took ▶ these pictures. This is a man's whisker six hours after being shaved with a safety razor.
1100 X—Johnson Wax Photo

This whisker was sheared off six hours earlier by ▶ an electric razor.
1500 X—Johnson Wax Photo

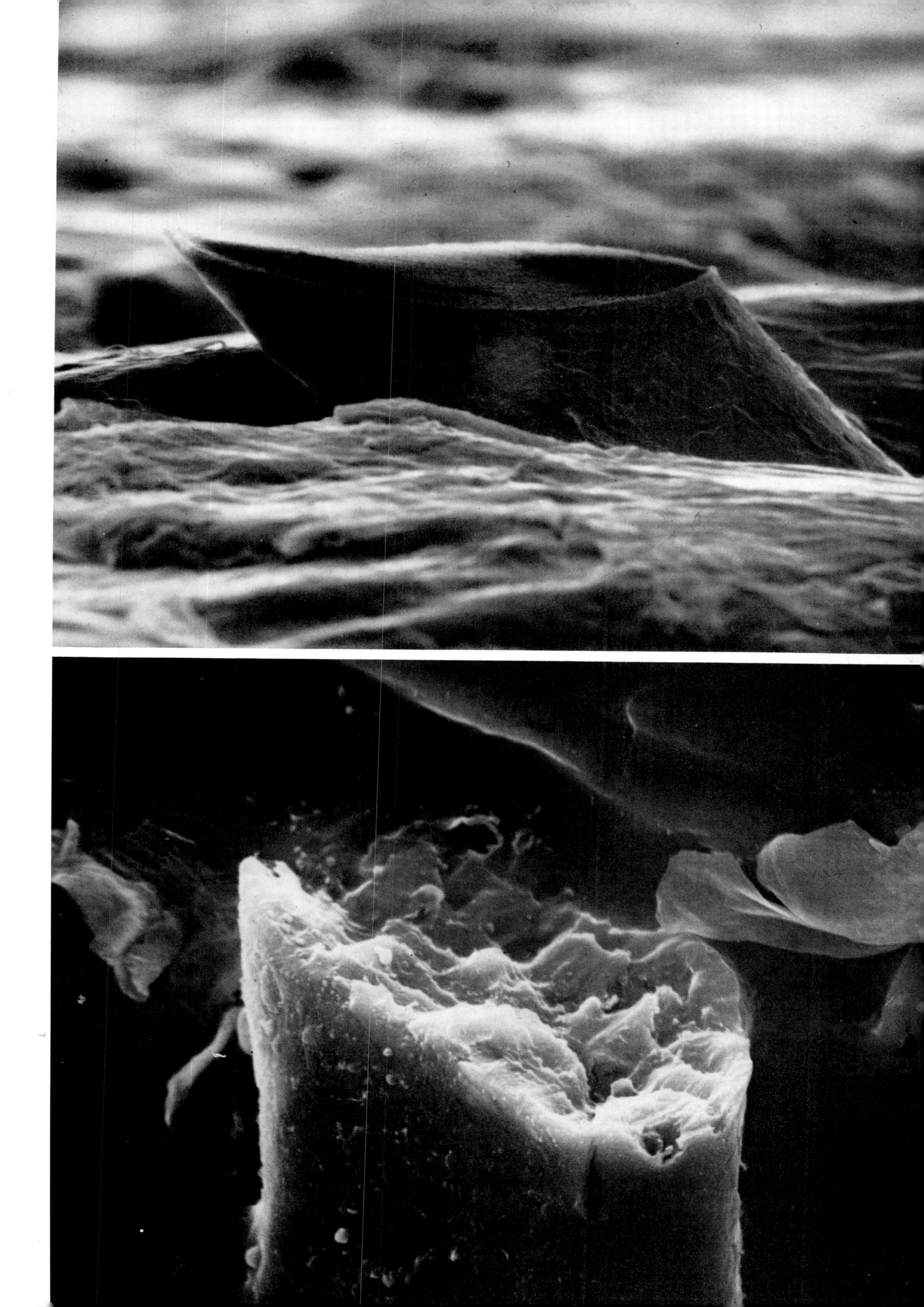

CHAPTER 5

the physical universe

Dust From Another World

When Apollo 11 astronauts Armstrong and Aldrin walked on the moon for the first time, they reported that the footing was slippery. When investigators on earth turned the scanning electron microsope on moon dust brought back by the astronauts, they found out why. The picture on the left shows a tiny, delicate spherule, magnified 7000 times. Moon dust contains huge numbers of tiny spheres; walking on it is almost like walking on ball bearings. Geologists theorize that the millions of meteorites that have smashed into the moon during its turbulent history threw huge amounts of molten materials miles above the surface. The fragments assumed a spherical shape, as do all fluids in free fall. Many of them solidified in that form before returning to the surface. The fragment on the right, magnified 3600 times, may be part of a larger spherical structure.

General Electric Research and Development Center

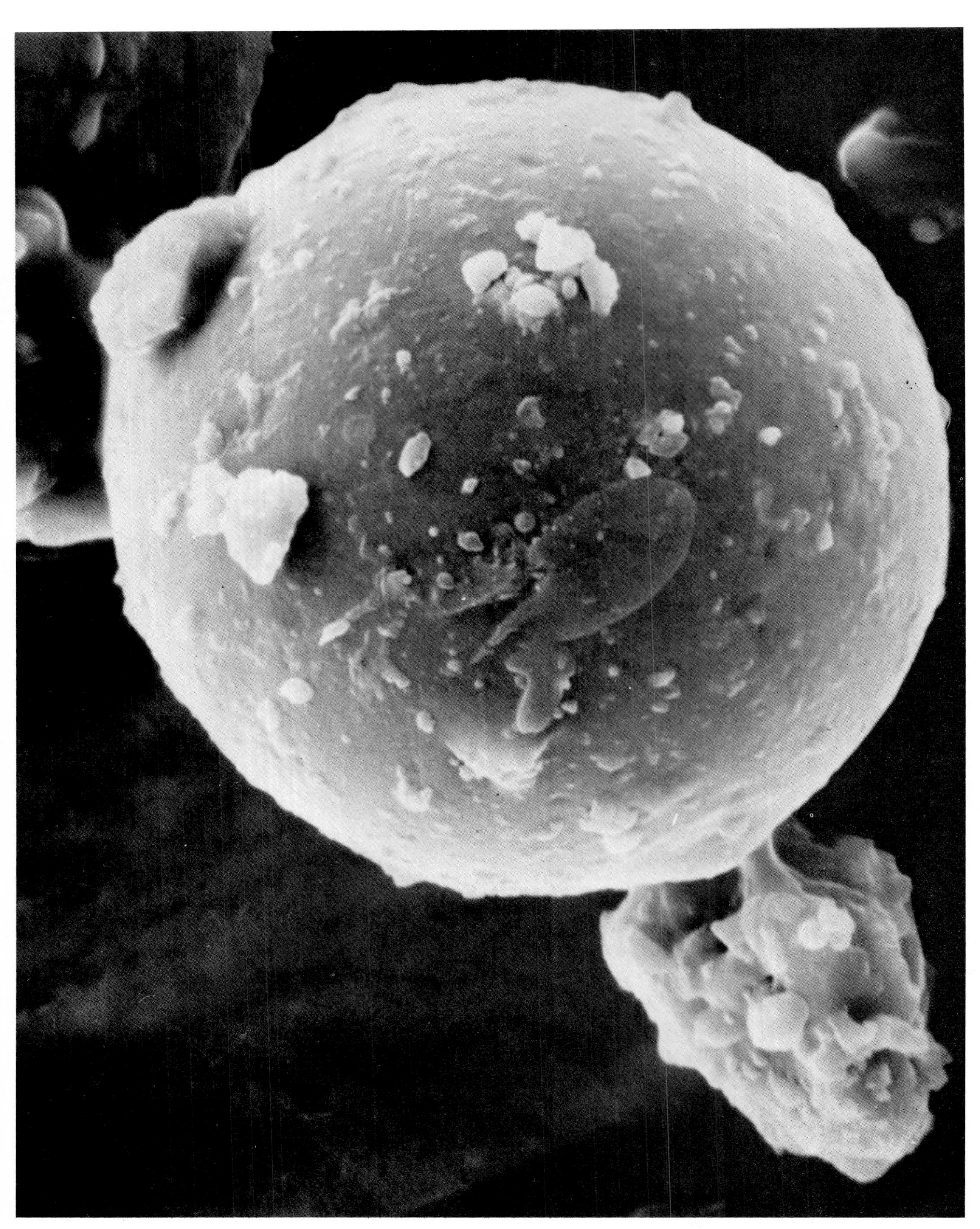

Minerals

One of the fields to benefit most from the extraordinary capabilities of the scanning electron microscope is metallurgy. Normal microscopes produce most of their images by transmitting energy—light or a stream of electrons—through a sample. But metals and most other minerals are opaque. Thus metallurgists usually have had to go through the tedious process of making minute and perfect molds or replicas of the surfaces of minerals to be able to view them. The scanning instrument has changed that, and in the process produced the spectacular photographs that follow.

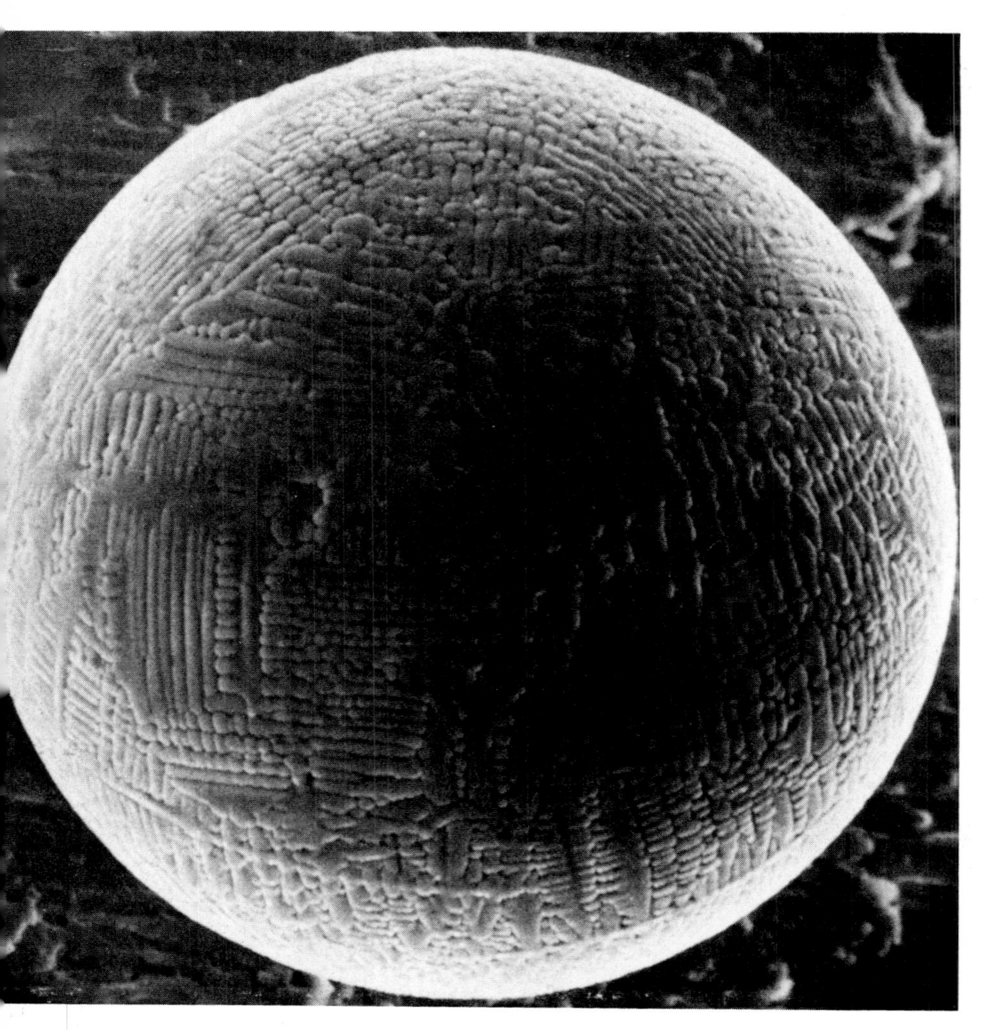

A striking new metal fabricating technique makes use of powdered metal, which can be formed into complex shapes and then hardened into strong, rigid structures. This is an extreme magnification of a particle of Astralloy, an alloy used in powdered-metal manufacturing processes. The magnificently regular crystalline structure of the particle is evident.
780 X—IIT Research Institute

Zinc orthosilicate ►
13,000 X—IIT Research Institute

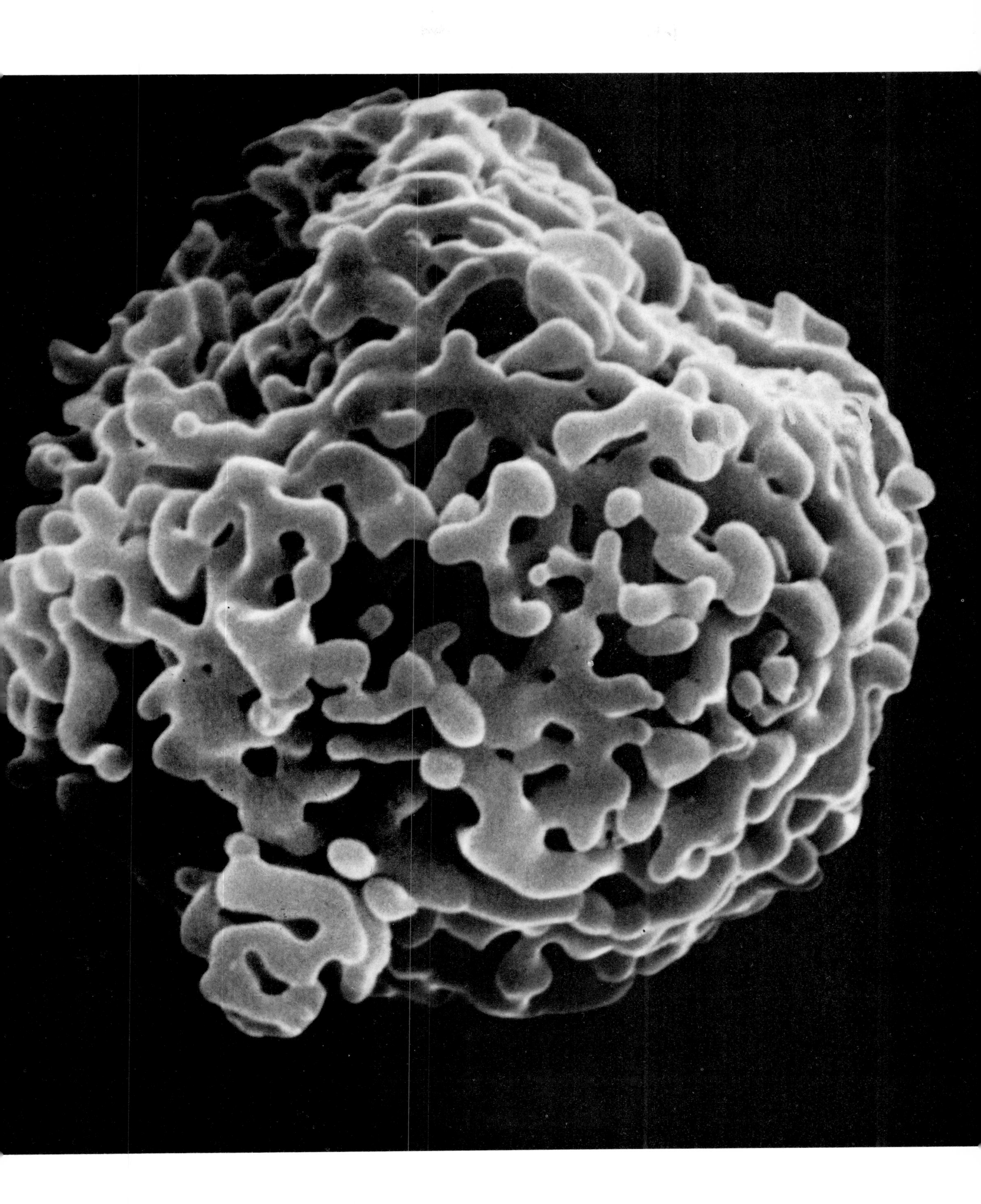

When certain metals and oxides are combined under certain conditions, the metal can form itself into slender fibers arranged in stunning regularity within the oxide matrix. These are shafts of tungsten that took shape in uranium dioxide. As many as ten million of them grow in a single square centimeter. The strange formation with circular areas free of fibers may result from supercooling of the liquid mixture during the growth process.
750 X—Dr. R. J. Gerdes, Georgia Institute of Technology

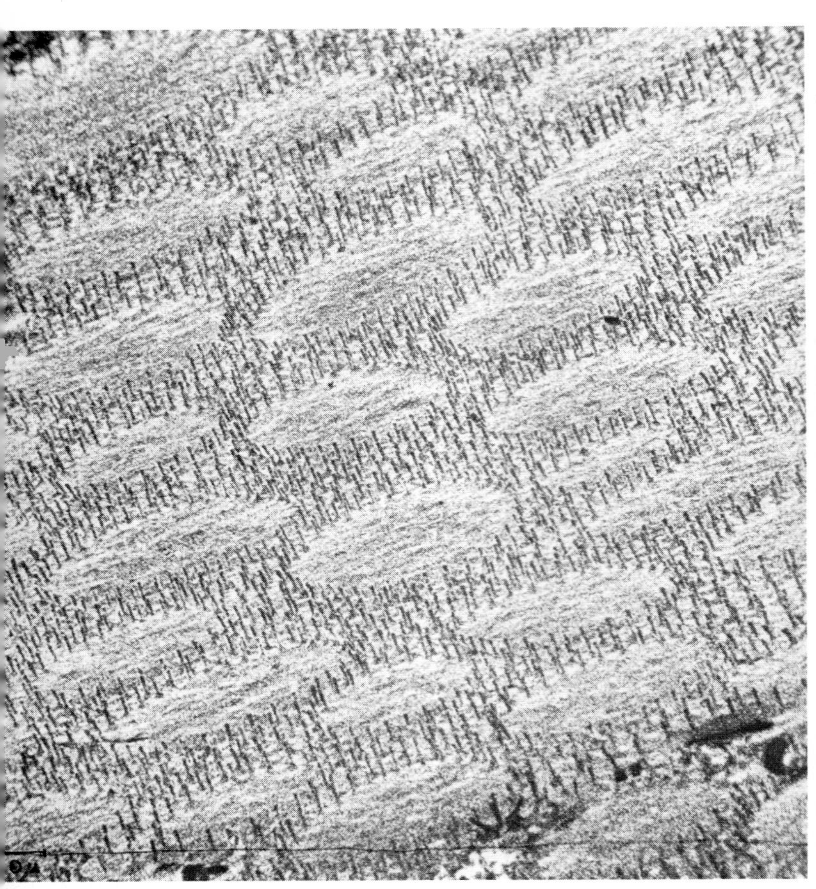

A closer view shows the regular construction of each tungsten fiber. The pointed end, formed by the facets, may be useful as an emitter in certain kinds of electronic circuits. The fibers are about 4 microns long—that's four thousandths of a millimeter or .0001576 inches.
9500 X—Dr. R. J. Gerdes, Georgia Institute of Technology

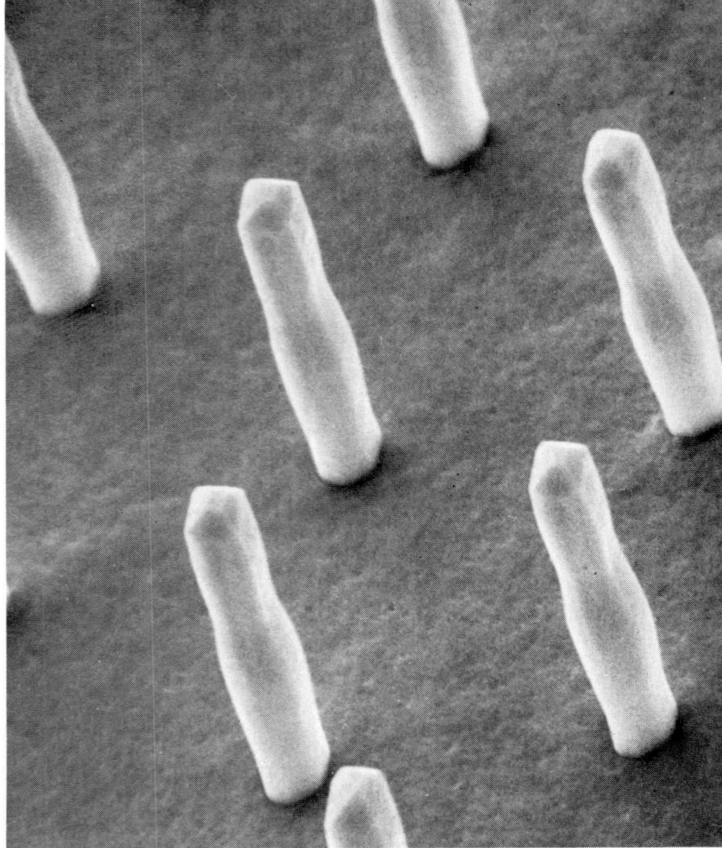

These beautiful patterns are called "Chinese script" characters by metallurgists. The sample is a nickel-based superalloy; the light characters are carbide particles within the alloy.
3150 X—Richard F. La Croix, Pratt & Whitney Aircraft

In this sample the carbide particles form a continuous network running through a grain boundary in the alloy.
2150 X—Richard F. La Croix, Pratt & Whitney Aircraft

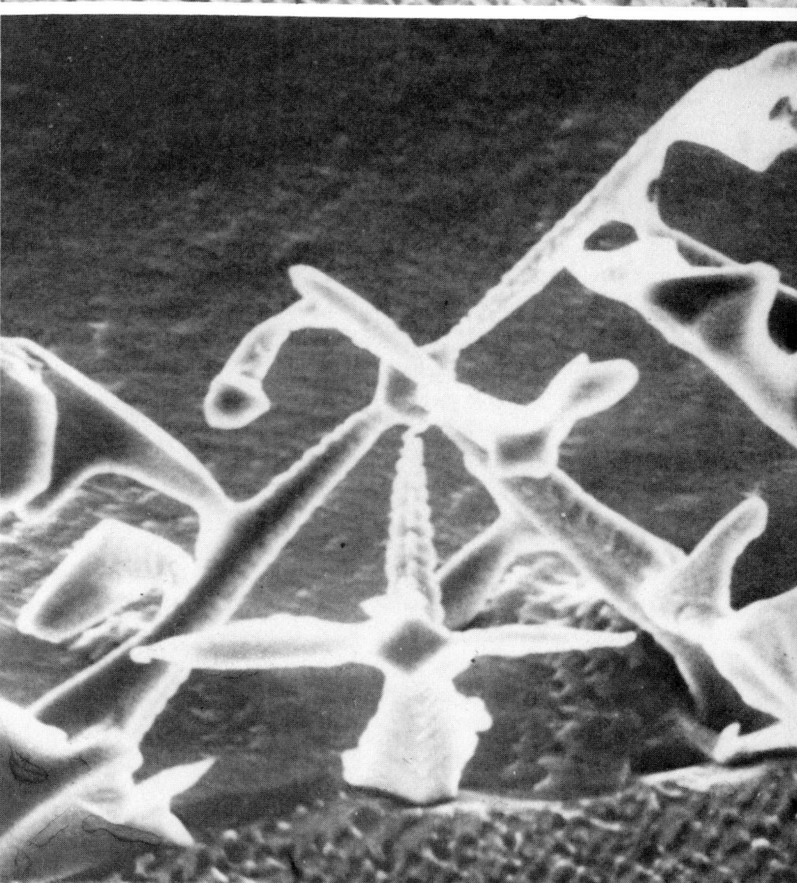

This pattern of carbides has formed into a light ► and fanciful structure.
2850 X—Richard F. La Croix, Pratt & Whitney Aircraft

The shape of any crystalline substance is determined by the shape of its constituent atoms, with each material assuming its characteristic and unique form. These fern-like fantasies are zinc dendrites, or branchlike growths. Though lovely, they cause a great deal of trouble in certain plating processes where they interfere with the regularity of the coating. Thus industrial processors go to great lengths to suppress their formation.
30 X—IIT Research Institute

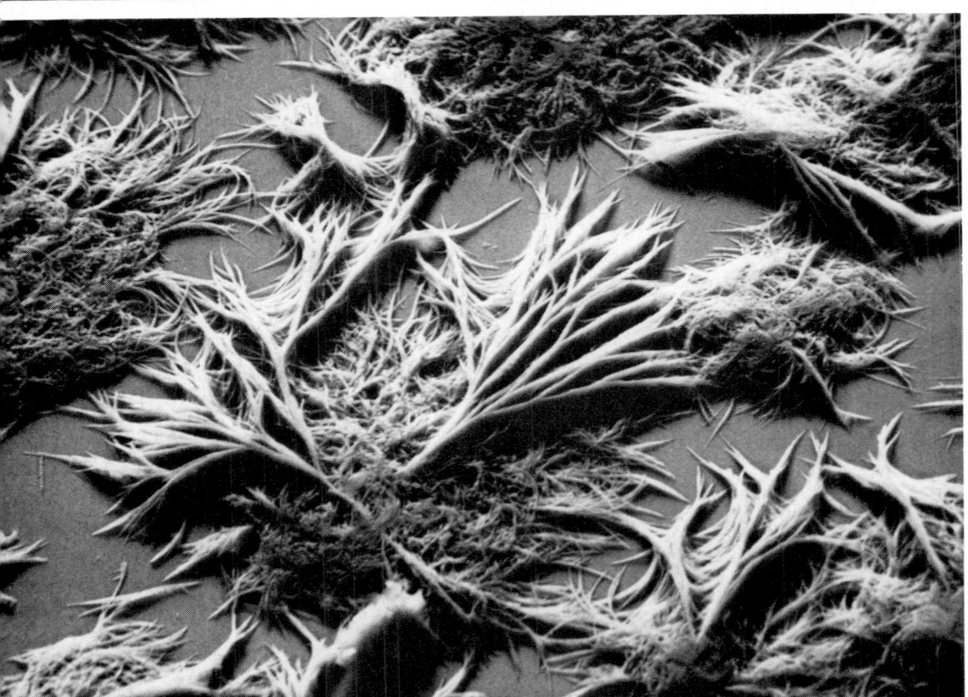

Sodium benzoate, a crystalline salt used principally as a food preservative.
820 X—Eastman Kodak Company, Industrial Laboratory

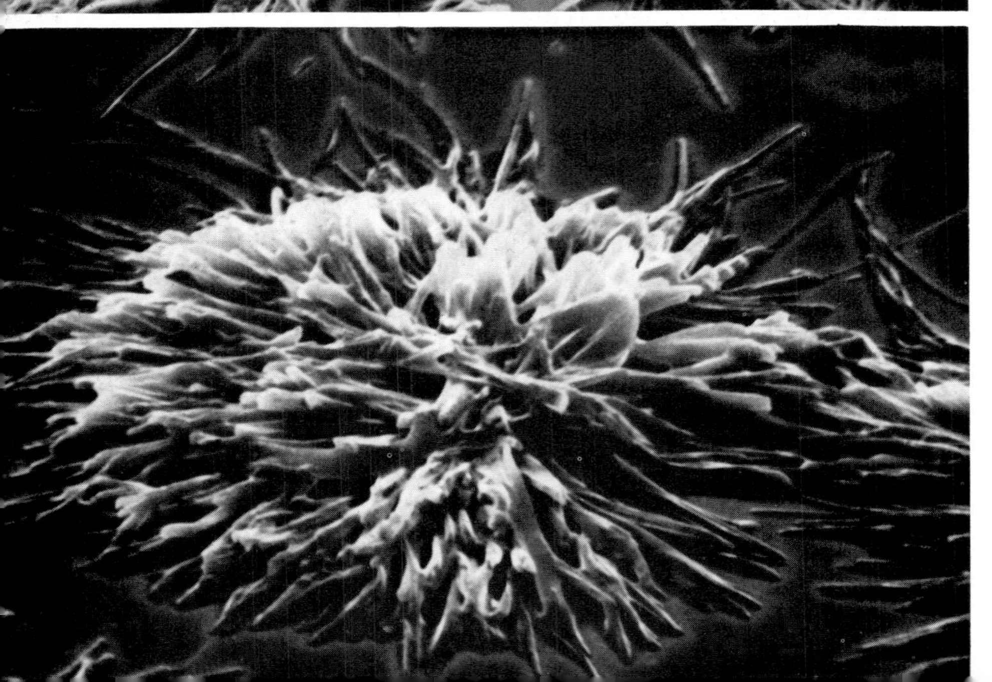

Mercury cyclohexane butyrate
3100 X—Eastman Kodak Company, Industrial Laboratory

The surface of chemically cleaned aluminum. The particles are aluminum oxide.
2650 X—IIT Research Institute

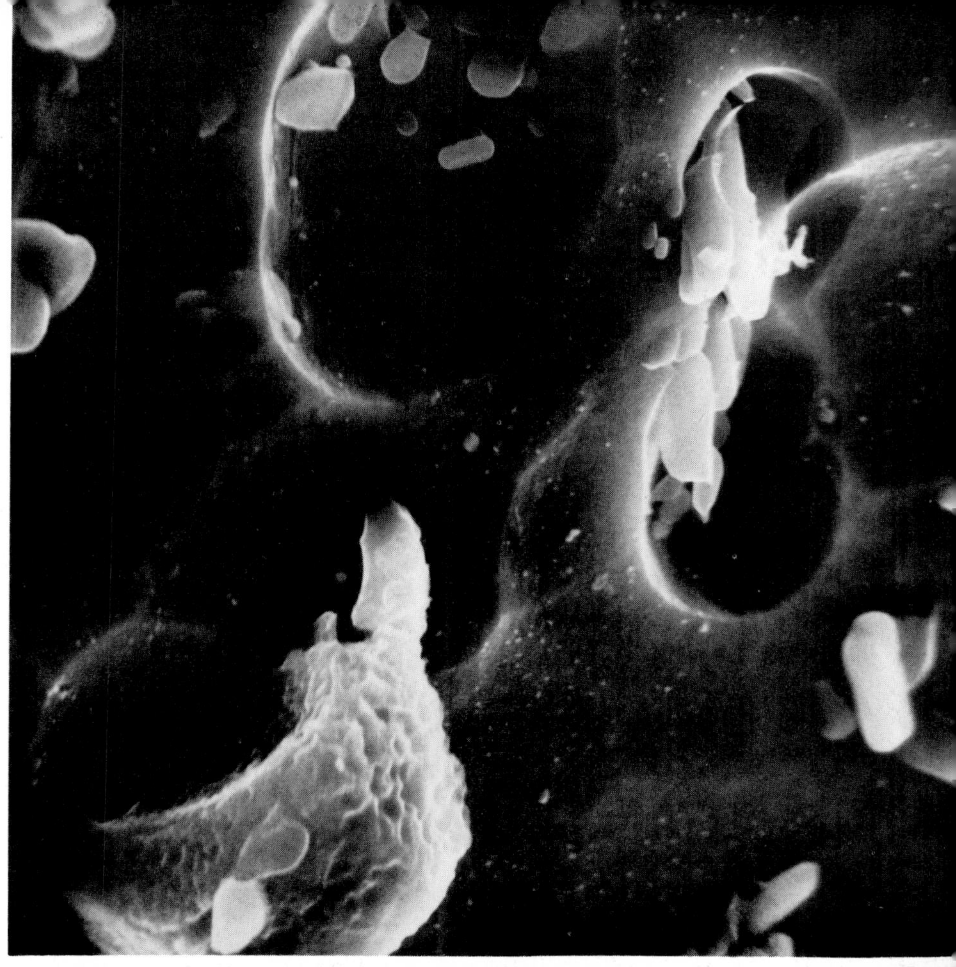

Corrosion pits on stainless steel
3800 X—Richard Turnage, Advanced Metals Research Corporation

Potassium acid phthalate
1600 X—Eastman Kodak Company, Industrial Laboratory

When antracene—a hydrocarbon distilled from coal tar—is subjected to heat and pressure, it changes to tiny spheres of carbon.
1550 X—Pyung Wha Whang, Materials Research Laboratory, The Pennsylvania State University

Deeply etched surface of a quenched and fractured melt of a semiconductor material used in electronic circuits.
3000 X—Takeshi Takamori, Solid State Science, The Pennsylvania State University

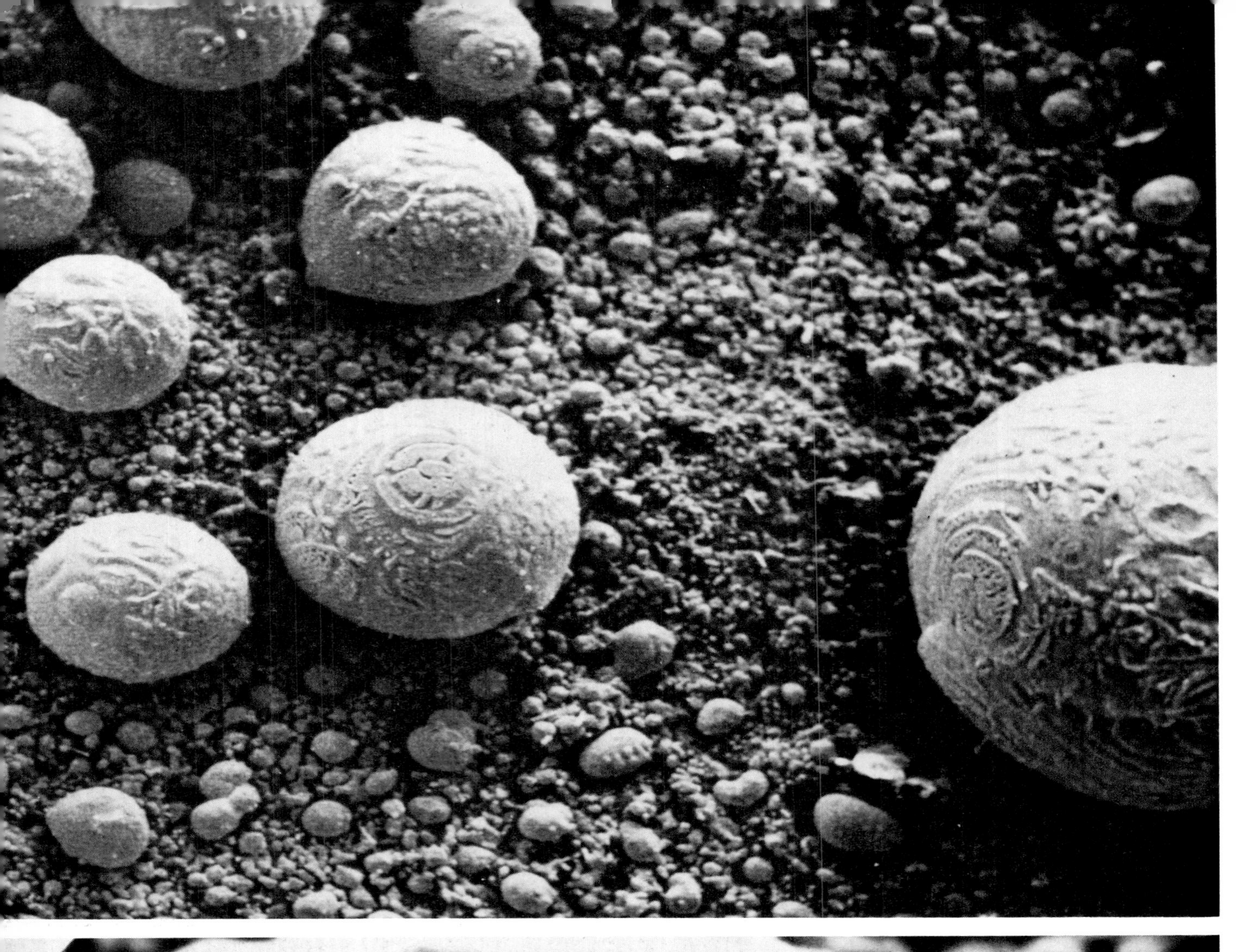

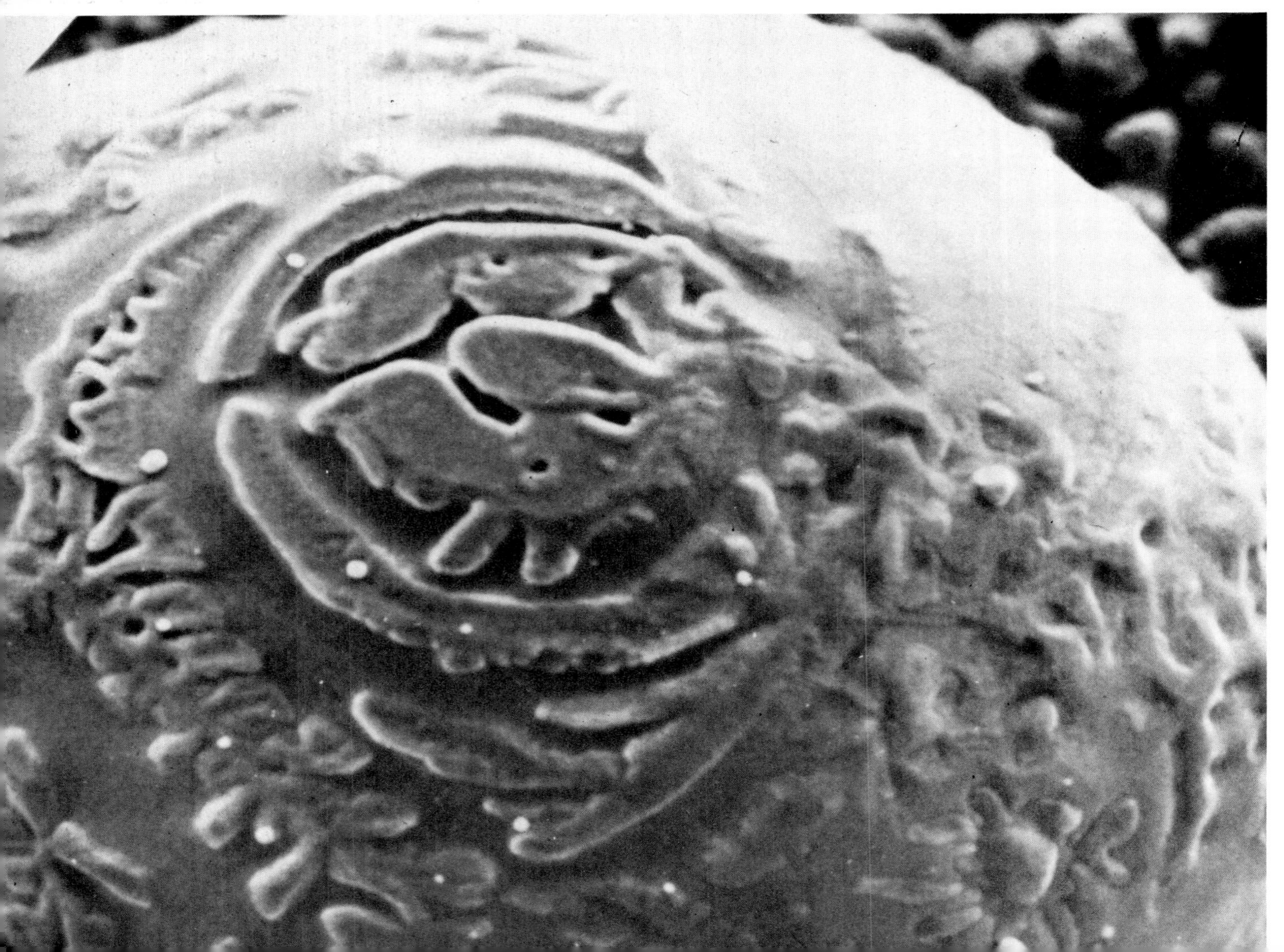

◄ **In the search for better alloys, metallurgists combine many metals. Here copper, iron, and tungsten were mixed and vaporized with extremely high temperatures. These patterned spheroids condensed from the metal vapor.**
600 X—Richard Turnage, Advanced Metals Research Corporation

◄ **A close-up (below left) shows the pattern in detail. The intricate design apparently results from dendritic growth, in which crystals accumulate in a branching, tree-like structure as they condense from a molten or vapor state.**
4000 X—Richard Turnage, Advanced Metals Research Corporation

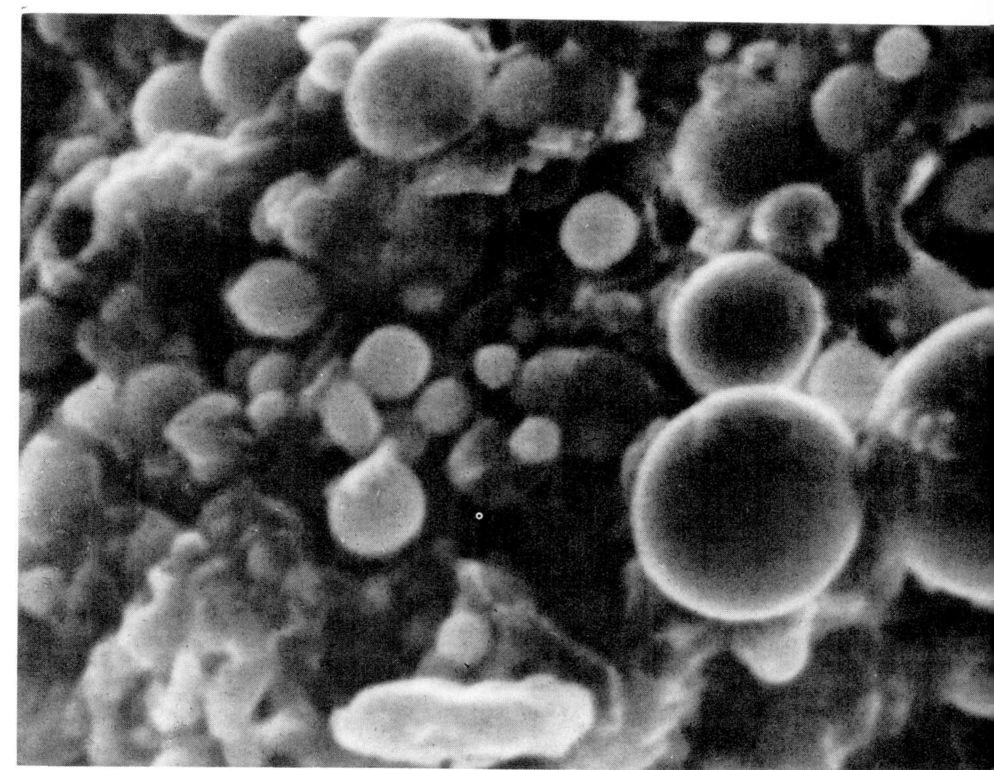

Smoke particles from a coal-burning steam plant
4000 X—Oak Ridge National Laboratory

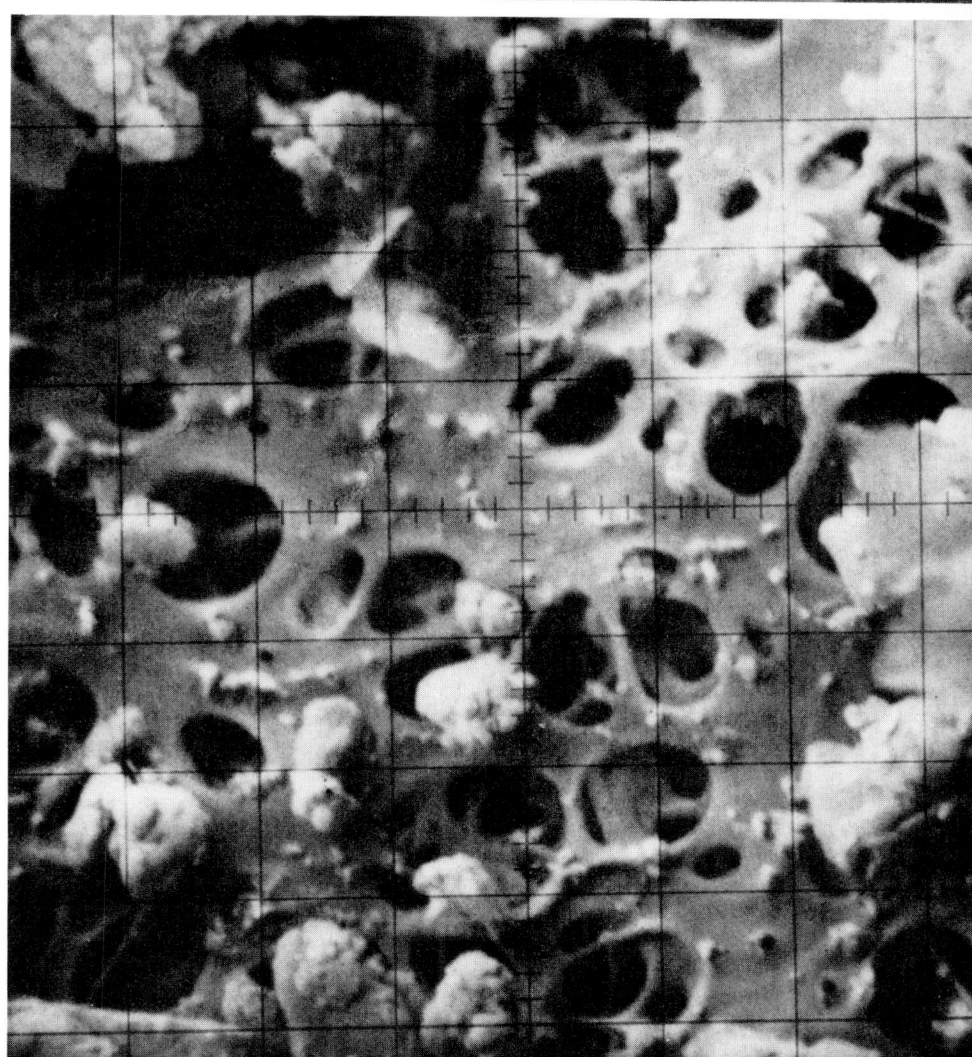

Dust particle caught on a filter. These bits of mineral, collected in a coal mine, can end up in miners' lungs.
10,000 X—Irene Rodgers, Philips Electronic Instruments

CHAPTER 6

specialized cells

Neurons

Of all the mysterious functions of life, no bodily process is so little known as that of the nervous system and the brain. "We don't understand the organization of even the simplest nerve units," says Professor Edwin R. Lewis of the electrical engineering department of the University of California at Berkeley. "For example, one particular nine-cell ganglion (nerve unit) from a lobster has been studied for 20 years, but its operation still isn't understood because it is not yet adequately mapped." Elements of the human nervous system are even less understood.

Using the scanning electron microscope, Dr. Lewis and his colleagues have developed a technique that promises to supply the missing information. Lewis, like other members of the bioengineering group he heads, took undergraduate work in biology, then switched to engineering to master the complicated physical problems involved in studying submicroscopic structure. Now their work seems to have paid off.

The remarkable photograph on this page is the first picture of what appears to be a synaptic knob—the crucial structure through which a nerve impulse is transmitted from one cell to another. Each knob appears to be connected to one or more other knobs, or to other nerve tissue. To get this picture, the research team carefully dissected a sea snail, *Aplysia californica*, known to have large and easily identified nerve cells in its abdomen. To prepare the specimens for microscopic study, they carefully pulled away the outer layer and subjected it to a multi-step process of soaking in various solutions and dehydration.

Under the microscope, this weird landscape appeared. By taking successive pictures and by using the scanning electron microscope's ability to twist, turn, and tilt the specimen in almost any way, they were able to map fibers and linkages from one cell to another. Thus is one of the great bottlenecks in neuroscience broken.

8000 X—Dr. Edwin R. Lewis, Dr. Yehoshua Y. Zeevi, Dr. Thomas E. Everhart, University of California, Berkeley

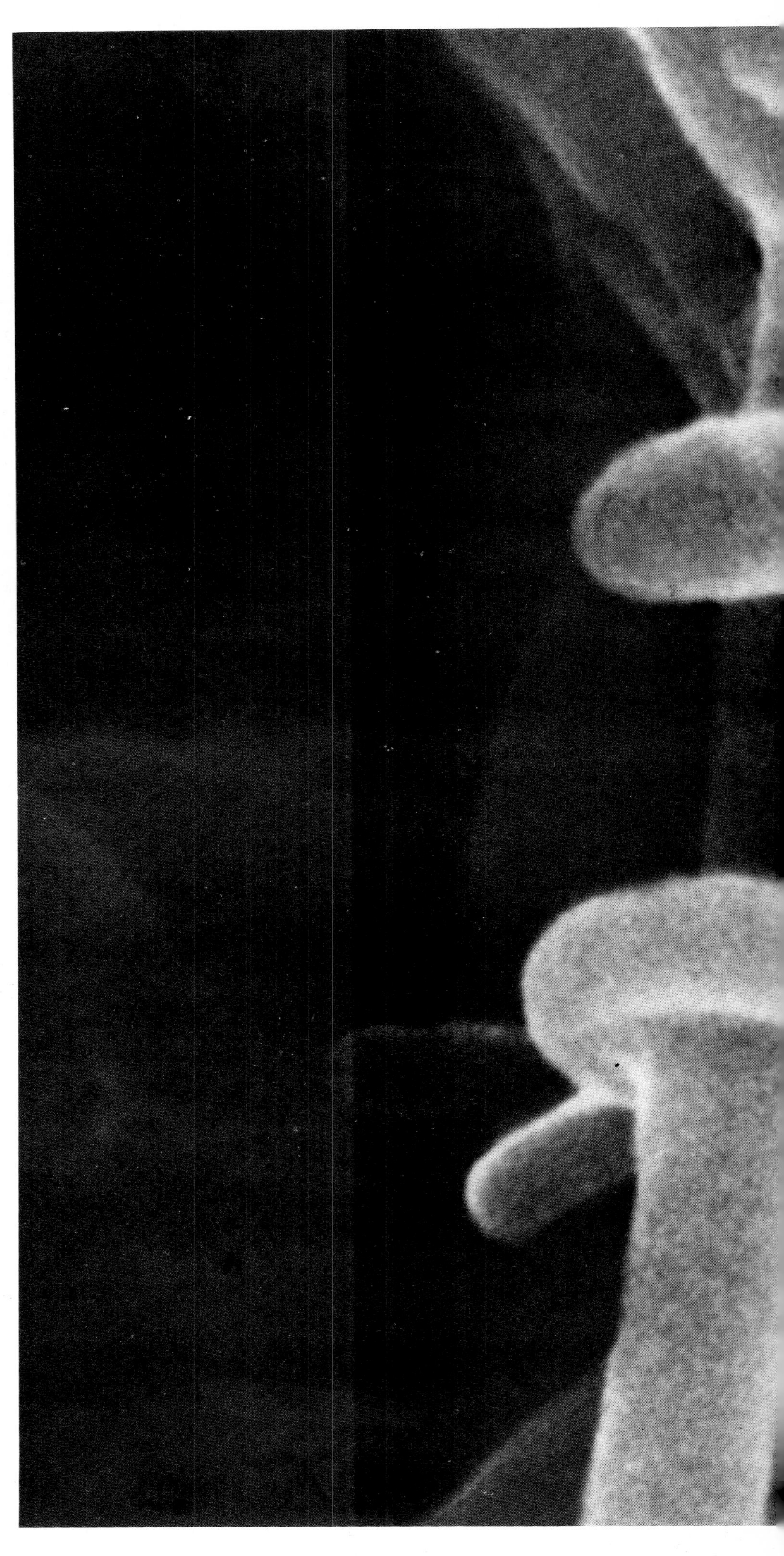

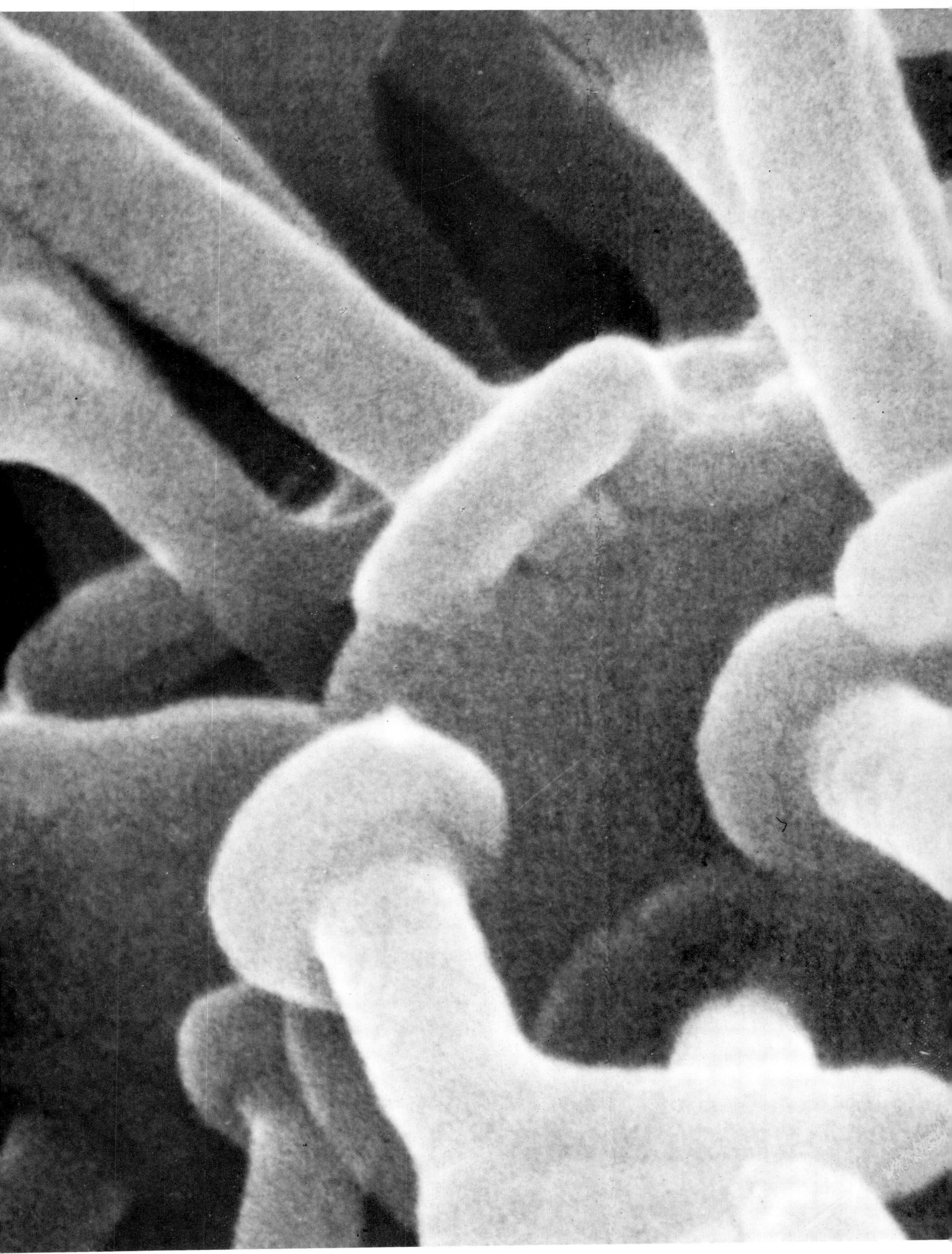

Cell Division

An unfertilized egg of the American leopard frog, *Rana pipiens*
52 X—Dr. L. M. Beidler, Florida State University

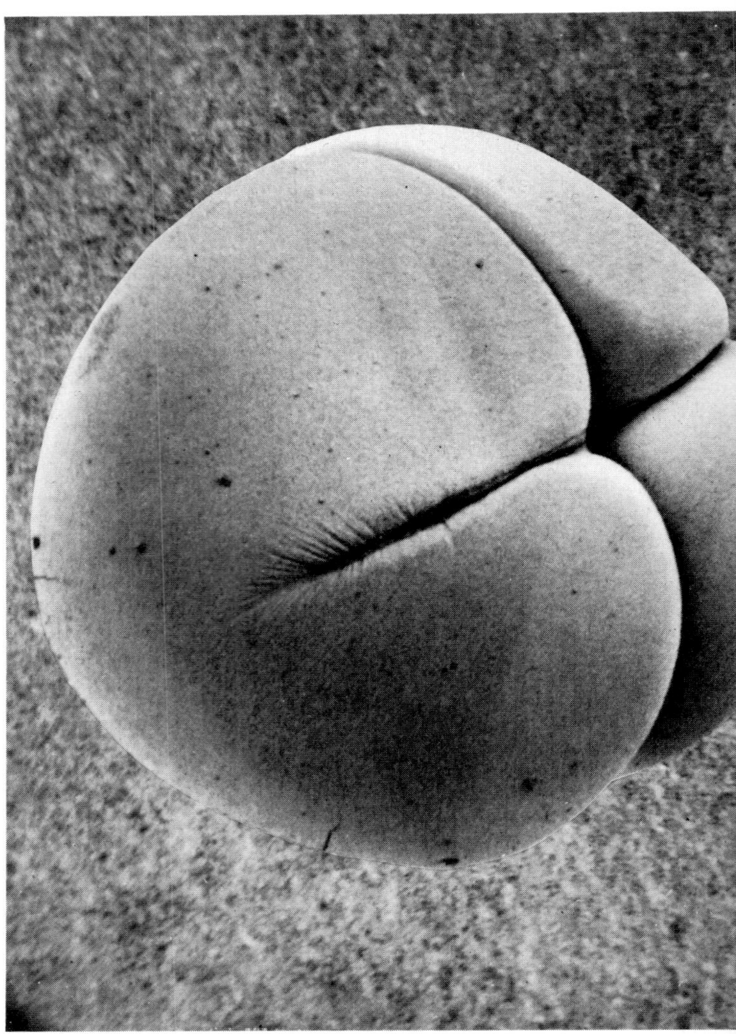

The egg is dividing into four cells
61 X—Dr. L. M. Beidler, Florida State University

One of the most remarkable processes in nature is the miraculous division of the fertilized egg into two cells, then two into four, four into eight, and so on, until millions or billions of cells of all types form a complete, functioning man, mouse, or frog. Since all cells during the early stages of cell division are identical, and since all cells in the body—be they skin or bone or liver cells—contain the same genetic information, what determines what any individual cell will become and what role it will play in the body? Despite much research, no one can yet answer this central question as to the roots of cell differentiation with any certainty. Yet it is clearly one of the key processes in life and growth. The unusual photographs that follow show the very earliest stages of the mysterious process of cell division up to the point where differentiation has begun.

Eight-celled stage
56 X—Dr. L. M. Beidler, Florida State University

Sixteen-celled stage of frog development
44 X—Dr. L. M. Beidler, Florida State University

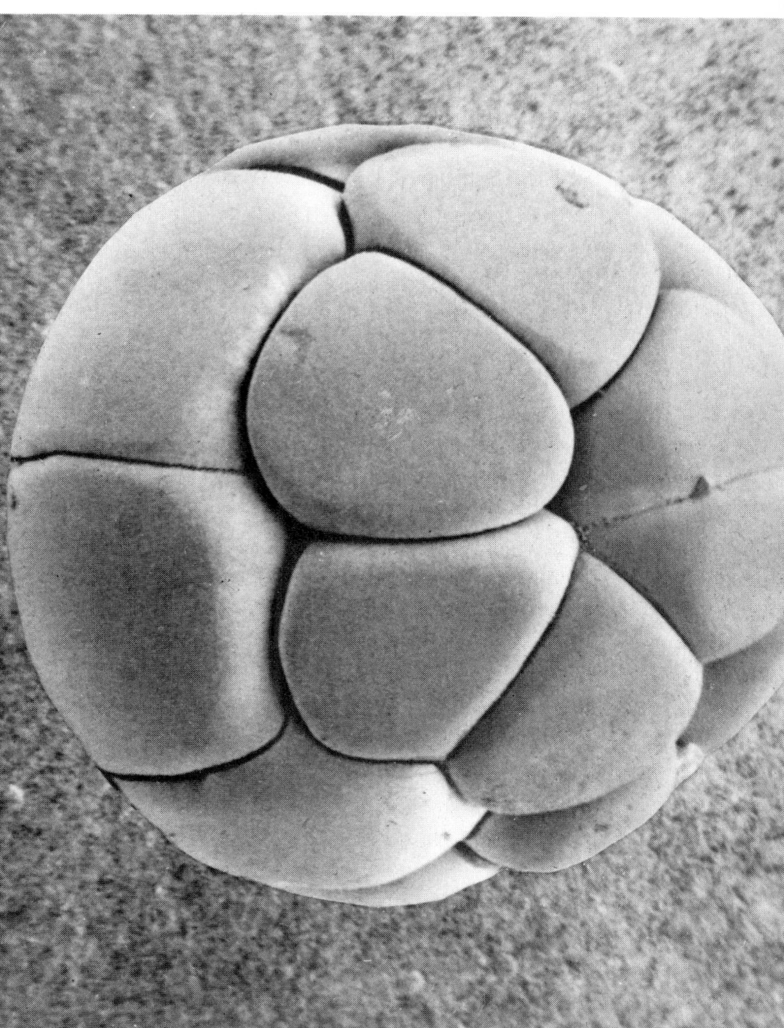

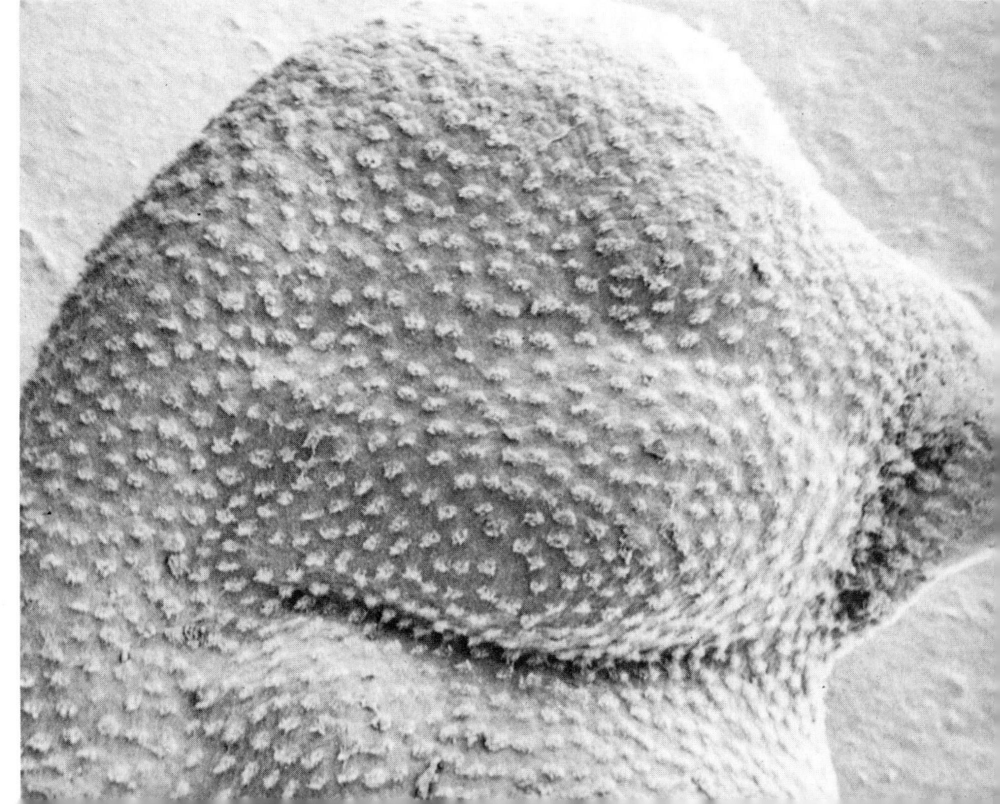

Ninety-four hours after fertilization, the original single cell has become thousands of cells and the embryo has begun the process of cell differentiation that will eventually result in some cells becoming eye cells, others muscle, others heart. This structure is the beginning of the tail bud.
84 X—Dr. L. M. Beidler, Florida State University

Blood Cells

Blood is the body's fluid of life. It performs many functions, but none is more important than its job as a carrier of oxygen to every cell in the body. This job is done by these almost doughnut shaped discs—the red blood cells.

9300 X—Dr. Patricia N. Farnsworth, Barnard College

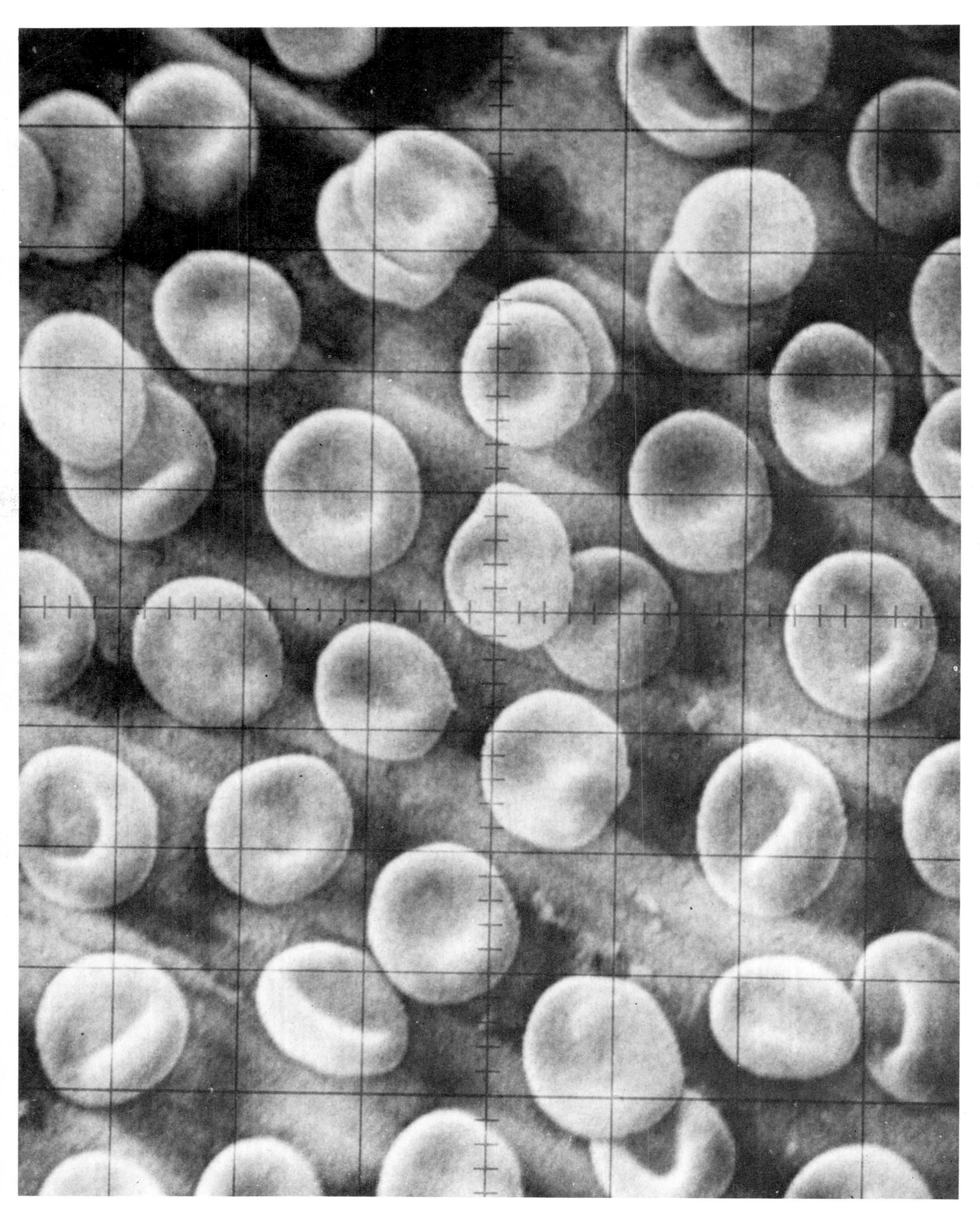

This striking picture shows the beginning of a vital process. The red blood cell is being enmeshed in a tangle of fibrin. This is the beginning of a blood clot, which occurs when the blood is exposed to air. Without the clotting function, even a scratch could be fatal.

26,000 X—Dr. Emil Bernstein, Eila Kairinen, Gillette Research Institute

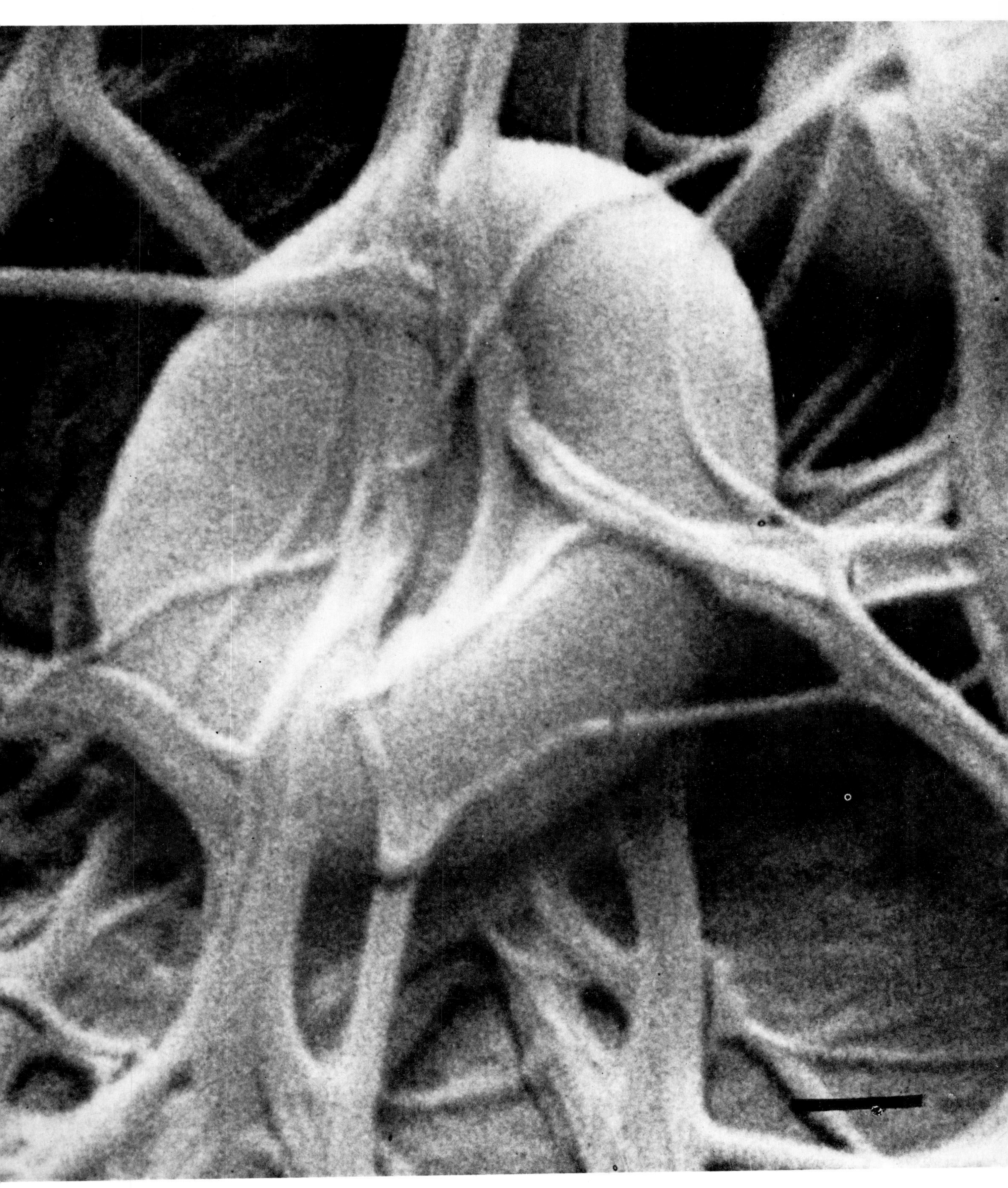

The disease that produces the grotesquely varied ► sickle cells is hereditary. Since it shortens its victims' lives and lessens their chance of living long enough to reproduce, the defective gene pool should tend to gradually disappear. Yet the opposite has happened; in some African tribes the sickle trait affects one out of two. The reason for this apparent failure of evolution is an extraordinary example of the marvelous complexity and subtlety of nature.

In Africa and Southern India, where a chance mutation countless centuries ago produced the defective hemoglobin that causes sickle cell anemia, malaria has wiped out large parts of the population repeatedly throughout history. For some reason, the sickle cells, though they eventually proved fatal, offered some immunity to malaria. And they prevented the high fevers that accompany it.

Thus those with the sickle cell trait, while eventually doomed, tended to survive into the age of childbearing in disproportionate numbers. Further, males with the sickle cells were not subject to malaria's high fevers, a single attack of which can depress sperm counts and sharply decrease fertility for many months. Thus sickle cell anemia victims were actually able to outbreed those not so affected.

13,000 X—Dr. Patricia N. Farnsworth, Barnard College

A normal blood cell has the familiar biconcave form—like a doughnut in which the hole fails to go all the way through. These cells are clearly not normal. They are produced by a mysterious disease called—for obvious reasons—sickle cell anemia. For presently unknown reasons, the disease causes red blood cells to change into weird forms—usually sickle or crescent shaped—such as these. Such cells clog capillaries, shutting off the supply of oxygen to affected regions. Victims suffer attacks of fever, weakness, severe pain. They usually die before the age of 30. The disease strikes blacks almost exclusively. One of every 10 blacks in this country carries the trait, but suffers no symptoms; two of every 1,000 have fatal anemia. There is no cure.

7000 X—Dr. Patricia N. Farnsworth, Barnard College

The peculiar shapes of red blood cells affected by ► sickle cell anemia are caused by a defective gene that produces an abnormality in the blood's hemoglobin, the protein pigment that gives blood its red color and carries oxygen to the tissues. But just why this causes the oddly characteristic deformation of cells is unknown. One investigator, Dr. Patricia N. Farnsworth, professor of biology at Barnard College, has been investigating sickled cells, and has, among other techniques, used the scanning electron microscope to study their shapes under many conditions.

25,000 X—Dr. Patricia N. Farnsworth, Barnard College

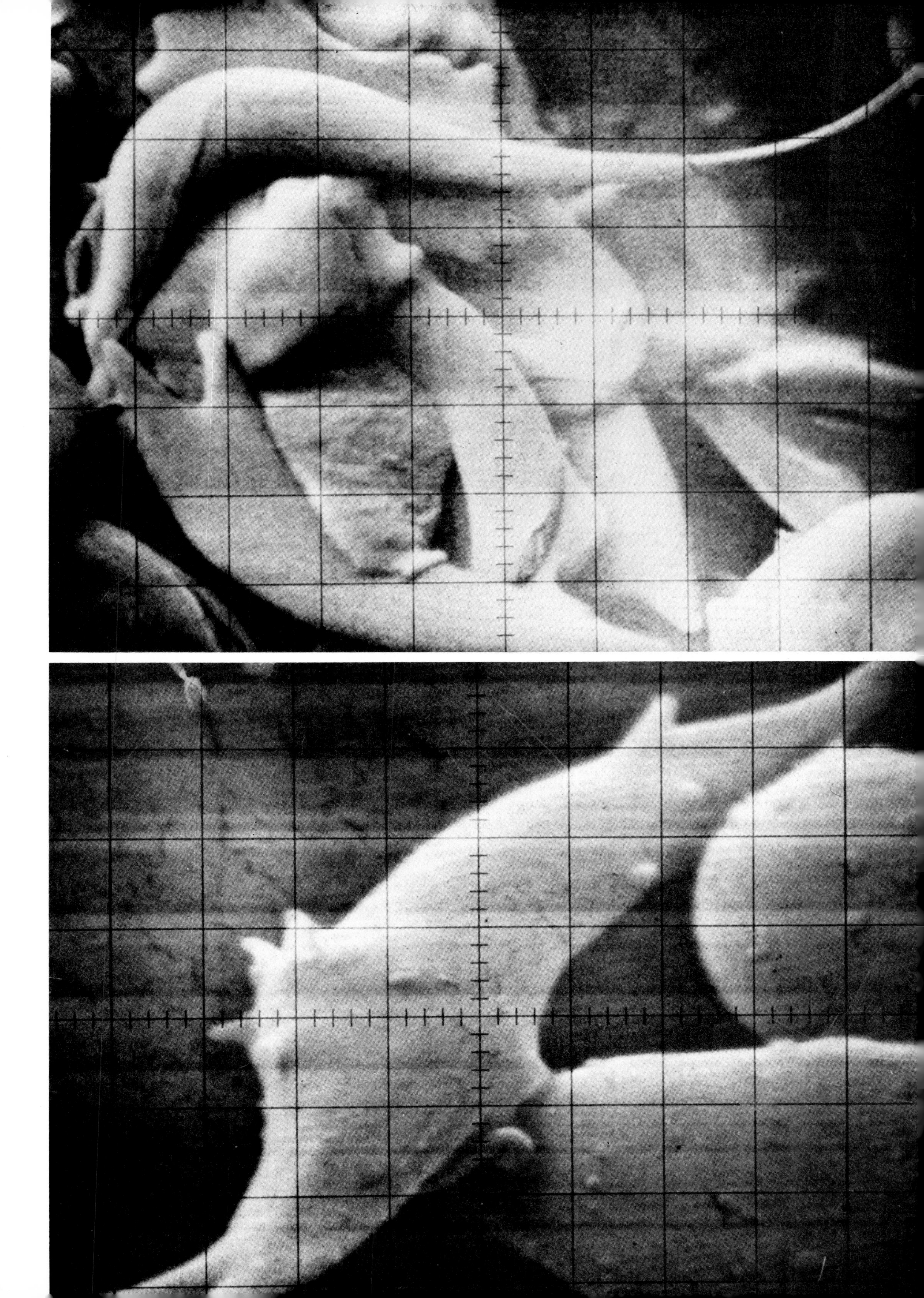

Until these studies, it was generally thought that the red blood cell's outer membrane was an inert sheath, playing no part in sickle cell anemia. But Dr. Farnsworth has produced evidence that it may perform a crucial function. During the course of experiments, she has treated cells with many substances that affect the outer membrane. These, for example, have been treated with an enzyme, called lysolecithin. It has caused selective shrinking of the center membrane leaving these porcupine-like protrusions.
5500 X—Dr. Patricia N. Farnsworth, Barnard College

These cells have been treated with albumen. They ► tend to become less biconcave, more spherical. Some display the characteristic "elves' hat" form shown by one in this cluster. But the important fact is this: both these cells and the porcupine-like ones are more resistant to sickling than normally shaped cells. Perhaps relatively simple drug or even dietary therapy that affects the red blood cell membrane—and thus its shape—could open an important new pathway to the treatment of a baffling malady.
26,000 X—Richard Turnage, Advanced Metals Research Corporation

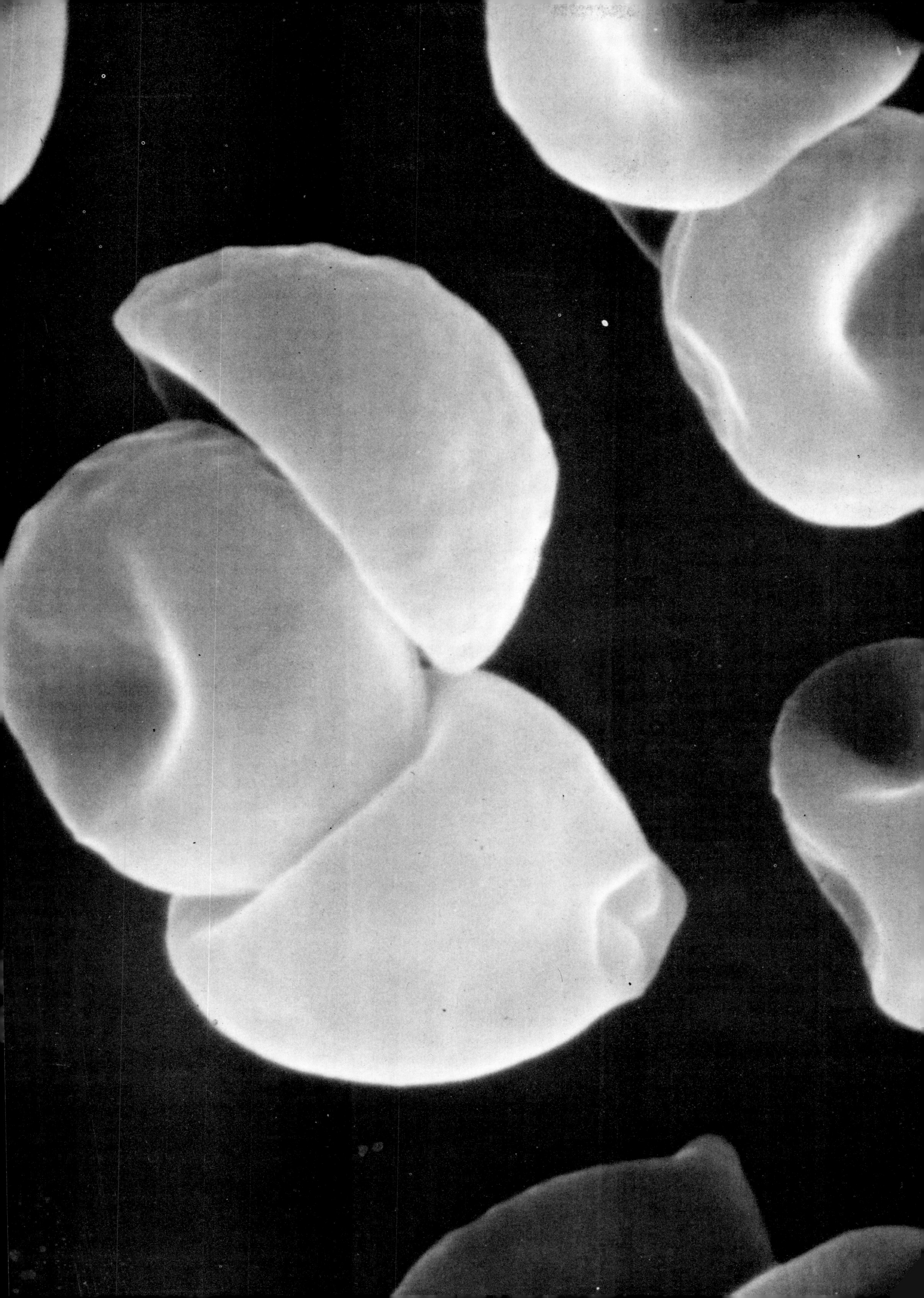

Sperm

Human spermatozoa
6000 X—Irene Rodgers, Philips Electronic Instruments

Life begins when a sperm cell from the male unites with the female's egg. Since the probability of a single microscopic sperm cell finding an egg in the female's reproductive tract would be slim, most species release vast numbers of sperm cells. With tails thrashing wildly, they blindly propel themselves randomly through the mucus that lines the tract. Sometimes, one bumps into the one or several eggs by accident and a new life starts. While sperm cells differ in appearance considerably from species to species, they all bear a family resemblance, and all display the prominent tail that plays such an important part in the beginning of life.

Spider crab sperm
27,000 X—Richard Turnage, Advanced Metals Research Corporation

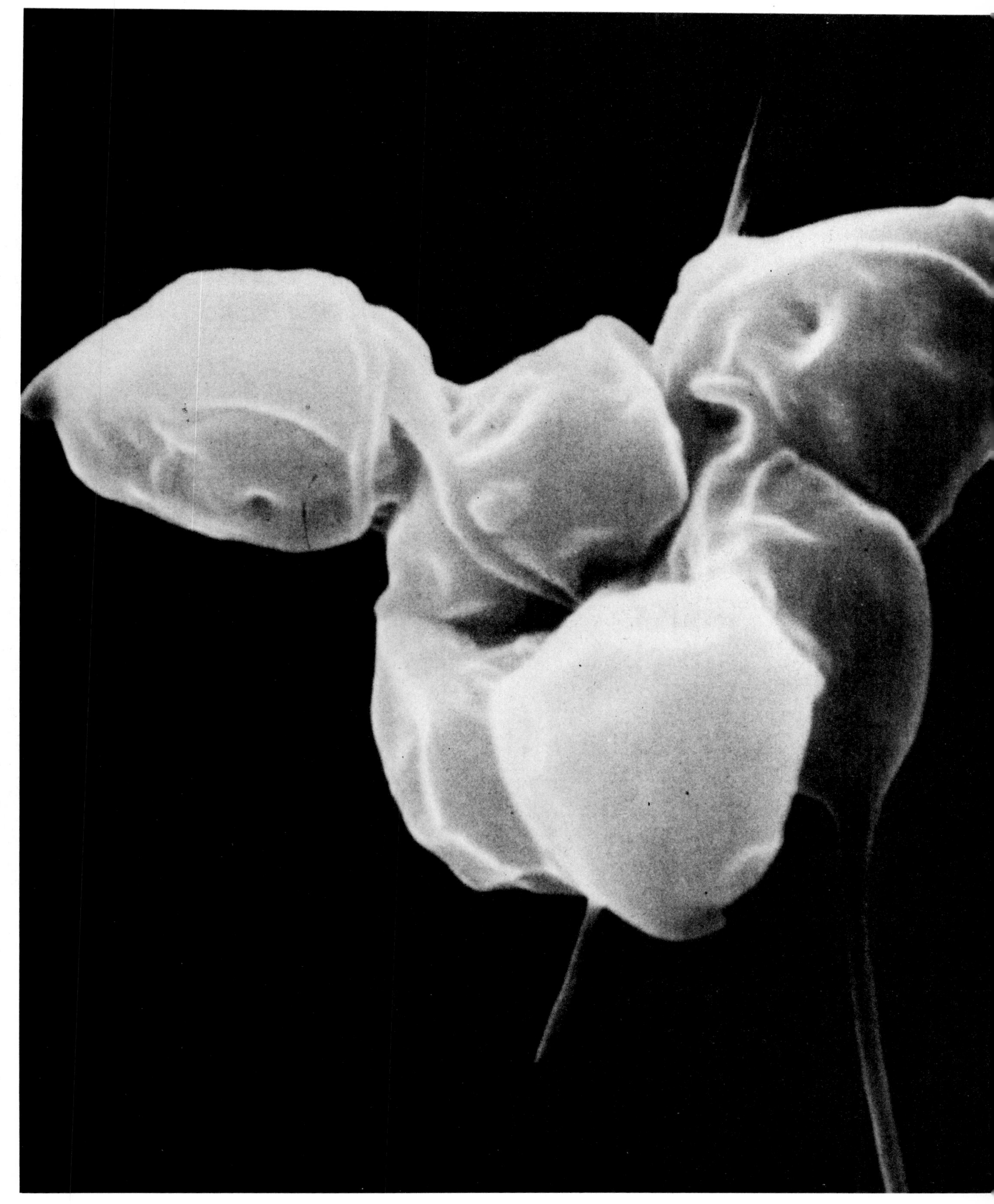

Rat sperm
Dr. Sarah A. Luse, College of Physicians and Surgeons, Columbia University

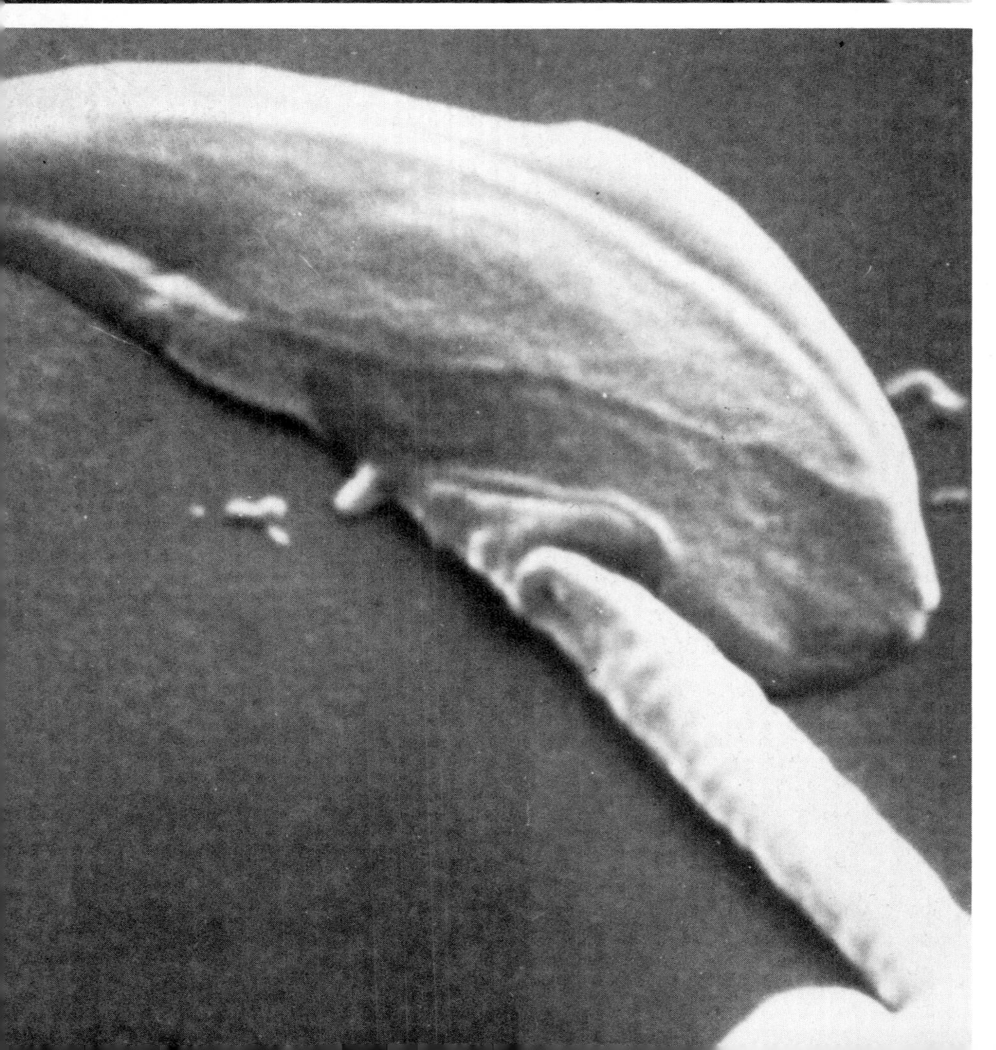

Mouse sperm
Dr. Sarah A. Luse, College of Physicians and Surgeons, Columbia University

Guinea pig sperm
Dr. Sarah A. Luse, College of Physicians and Surgeons, Columbia University

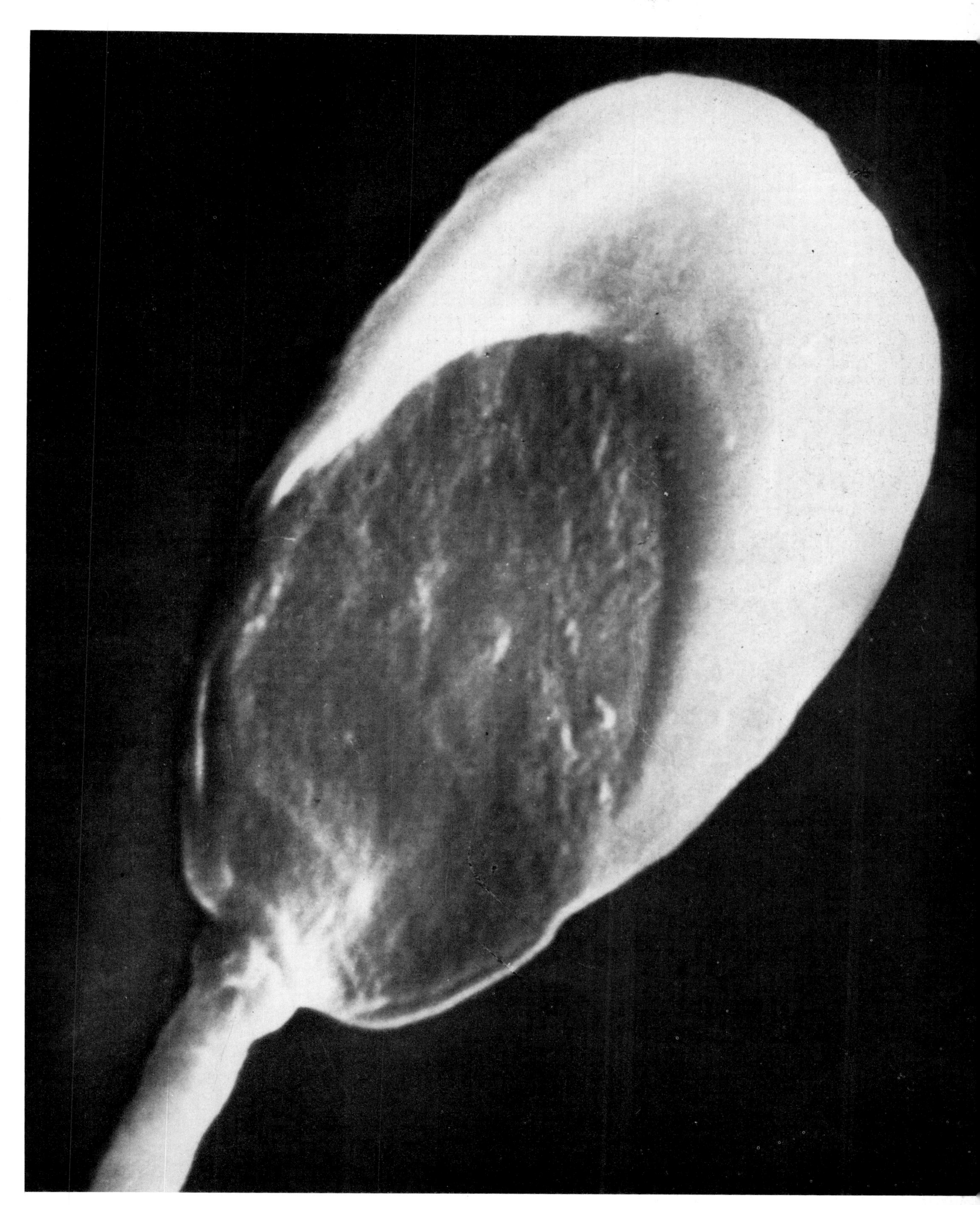

CHAPTER 7

minute life: animal, vegetable, and in-between

Bacteria Colony

Scientists have studied bacteria under the microscope for many years. But the unicellular creatures have most often been viewed on glass slides or under other unnatural conditions, so that it was impossible to tell how they live and function in nature.

The scanning electron microscope is now giving a new view of this micro environment. These boulder-like objects are sand grains gathered by gently pressing a sticky specimen stub from a scanning electron microscope onto a freshly exposed soil surface. Once the SEM operator locates an object such as a sand grain, simple controls on the instrument allow him to zoom easily for a closer look.

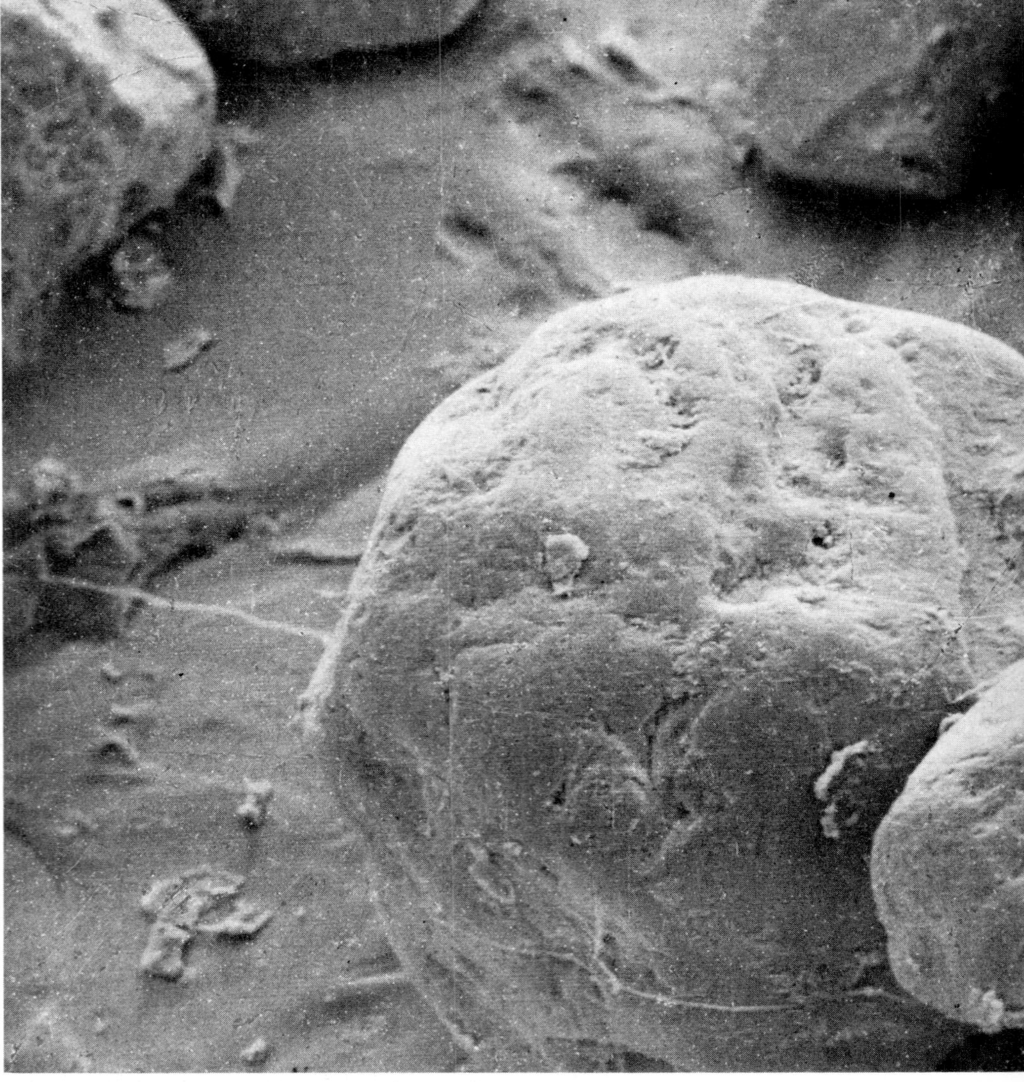

370 X—Dr. T. R. G. Gray, University of Liverpool

Two fuzzy-looking patches show up among the crater-strewn, debris-littered surface of the sand grain.
685 X—Dr. T. R. G. Gray, University of Liverpool

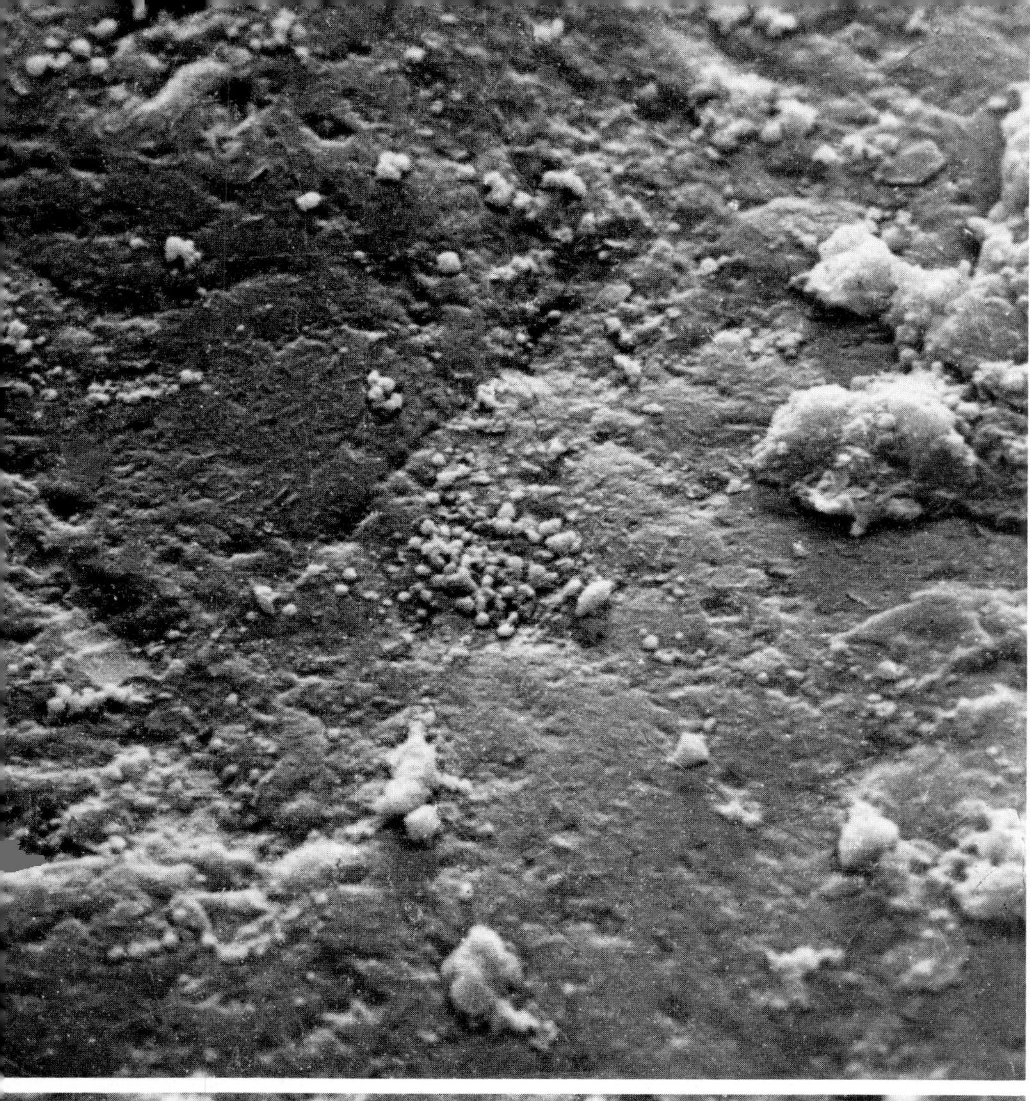

A still closer look reveals the cluster to be colonies of spherical bacteria. One cluster is at center, another near the left edge.
3750 X—Dr. T. R. G. Gray, University of Liverpool

A final zoom reveals many details about the structure, arrangement, and relationships between the nearly 100 bacteria in the colony. This simple technique will help solve many problems concerning the relation of micro-organisms to their environment.
8500 X—Dr. T. R. G. Gray, University of Liverpool

Unidentified rod-shaped bacillus on the surface of a tooth
3600 X—Dr. Emil Bernstein, Eila Kairinen, Gillette Research Institute

***Staphylococcus epidermidis*, a skin bacterium, on the hair of a guinea pig**
7000 X—Dr. Emil Bernstein, Eila Kairinen, Gillette Research Institute

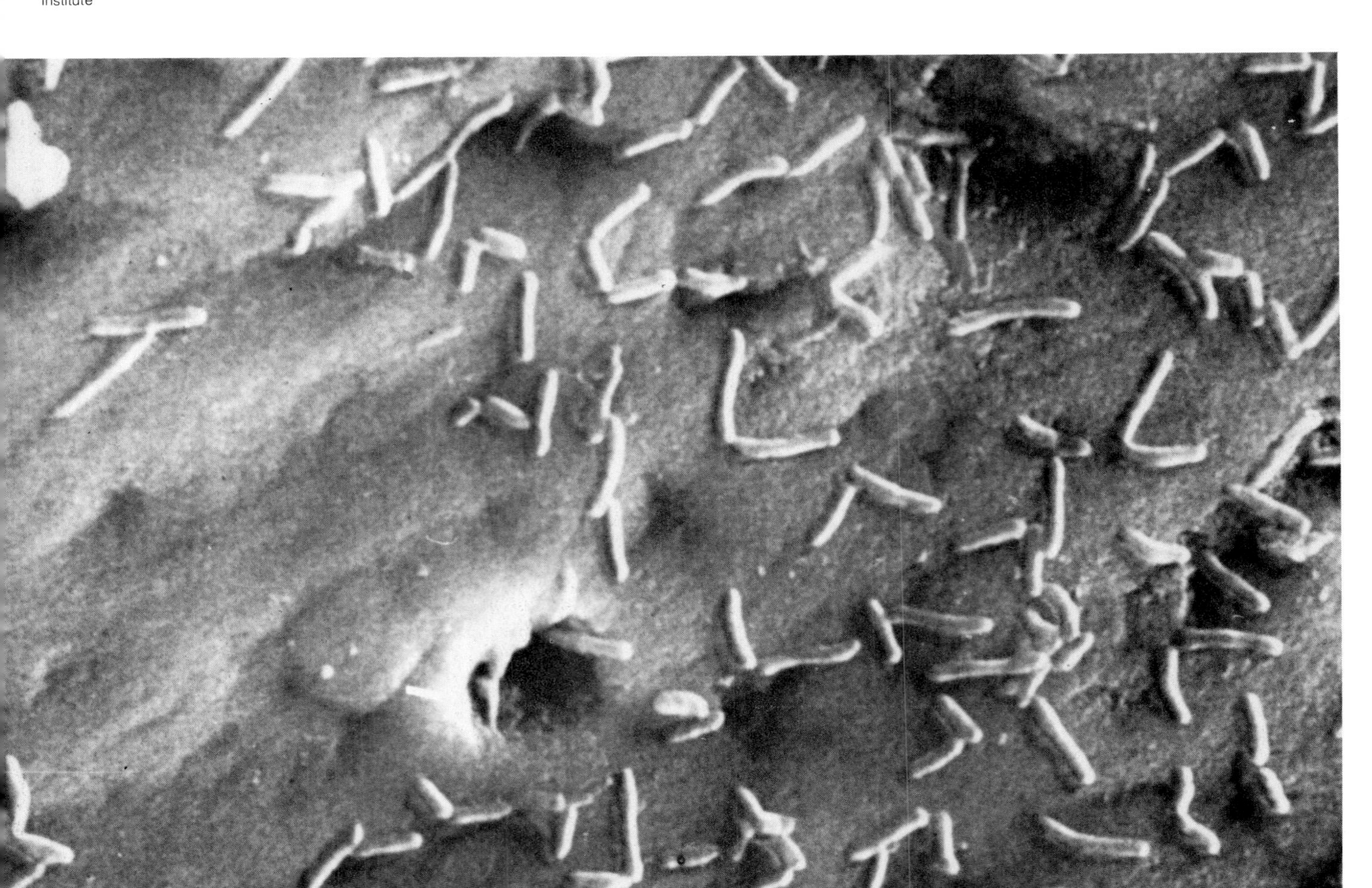

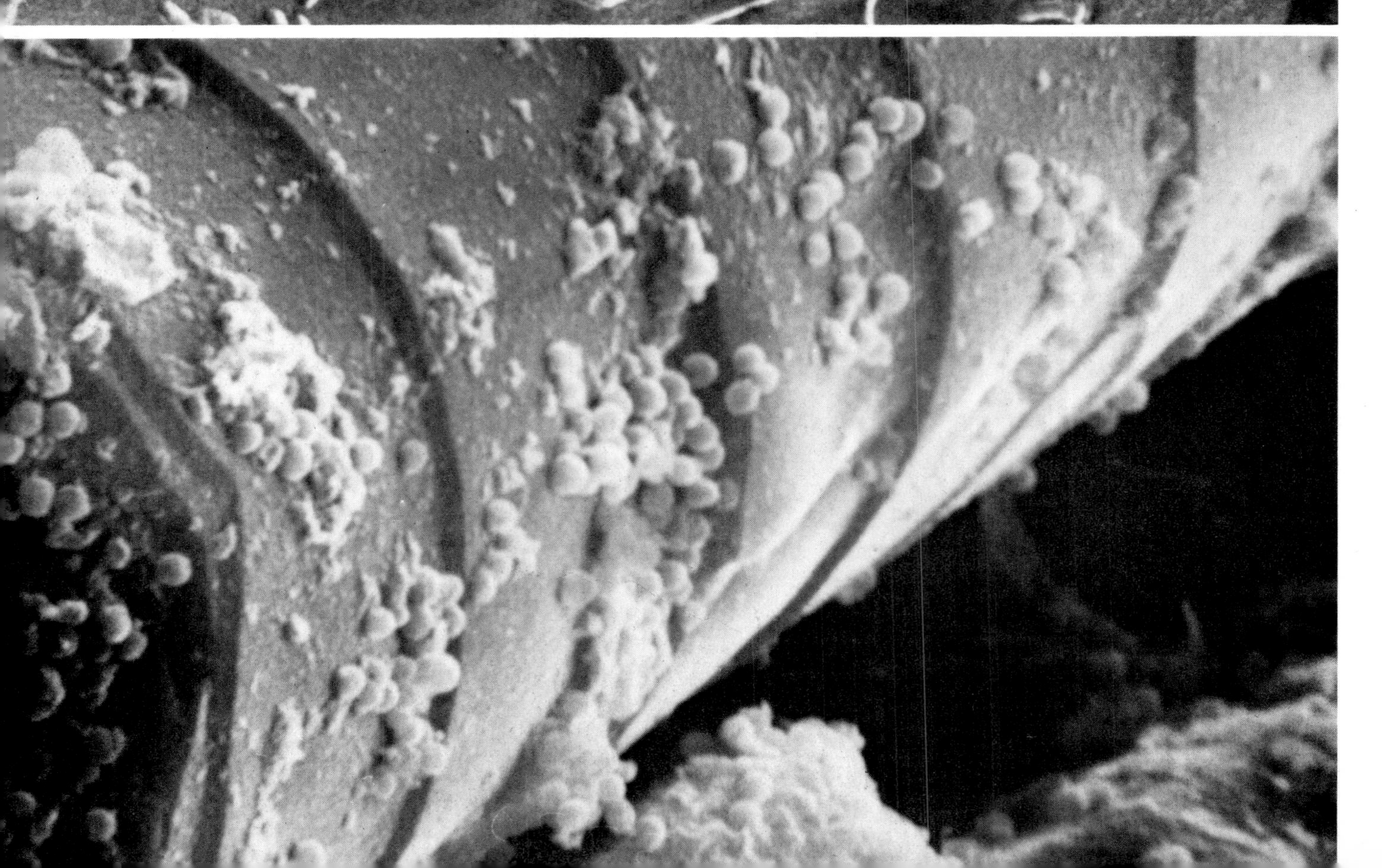

The same bacteria on guinea pig skin
4800 X—Dr. Emil Bernstein, Gillette Research Institute

***Bacillus megaterium* on the skin of a guinea pig**
5000 X—Dr. Emil Bernstein, Gillette Research Institute

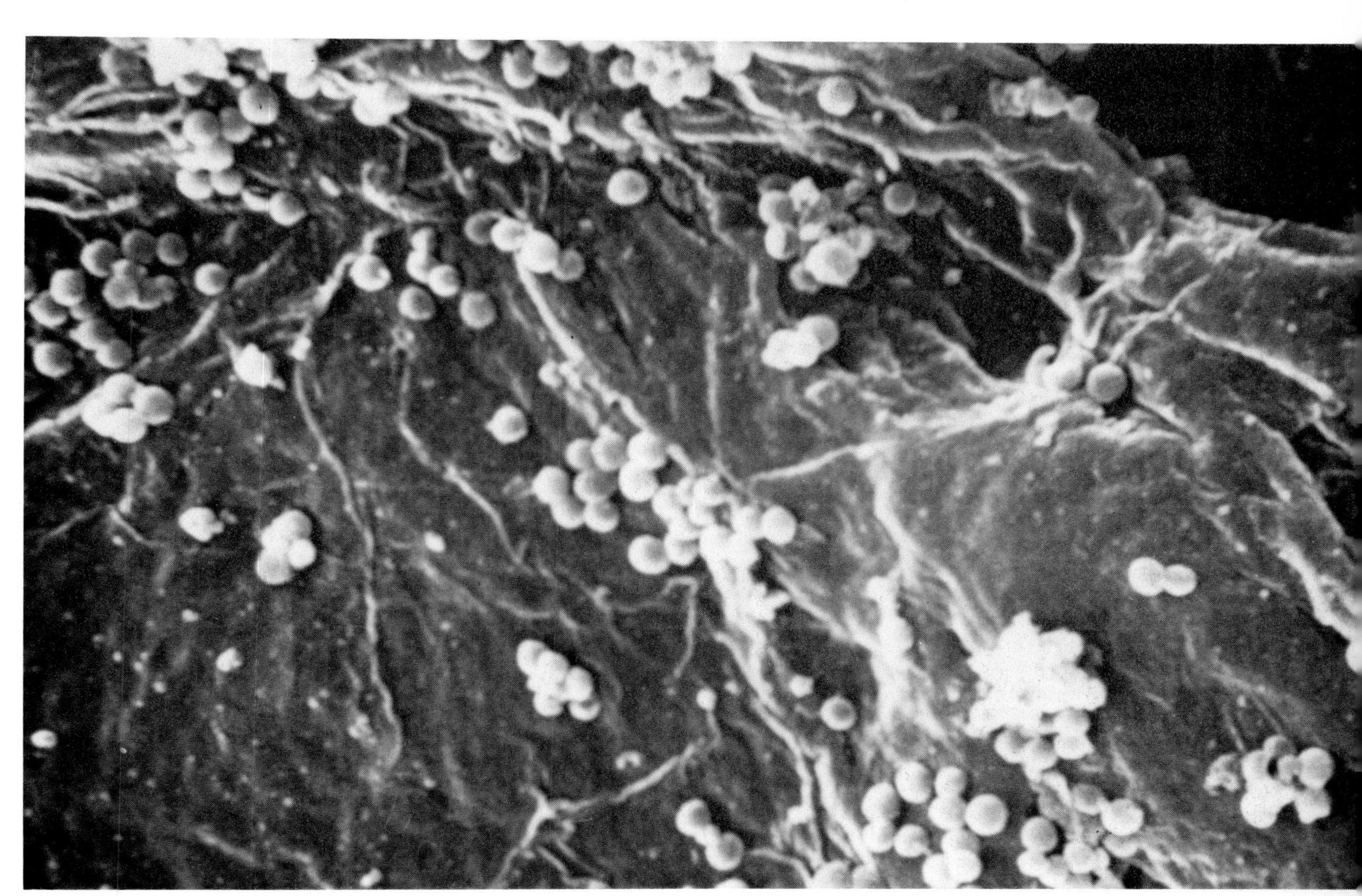

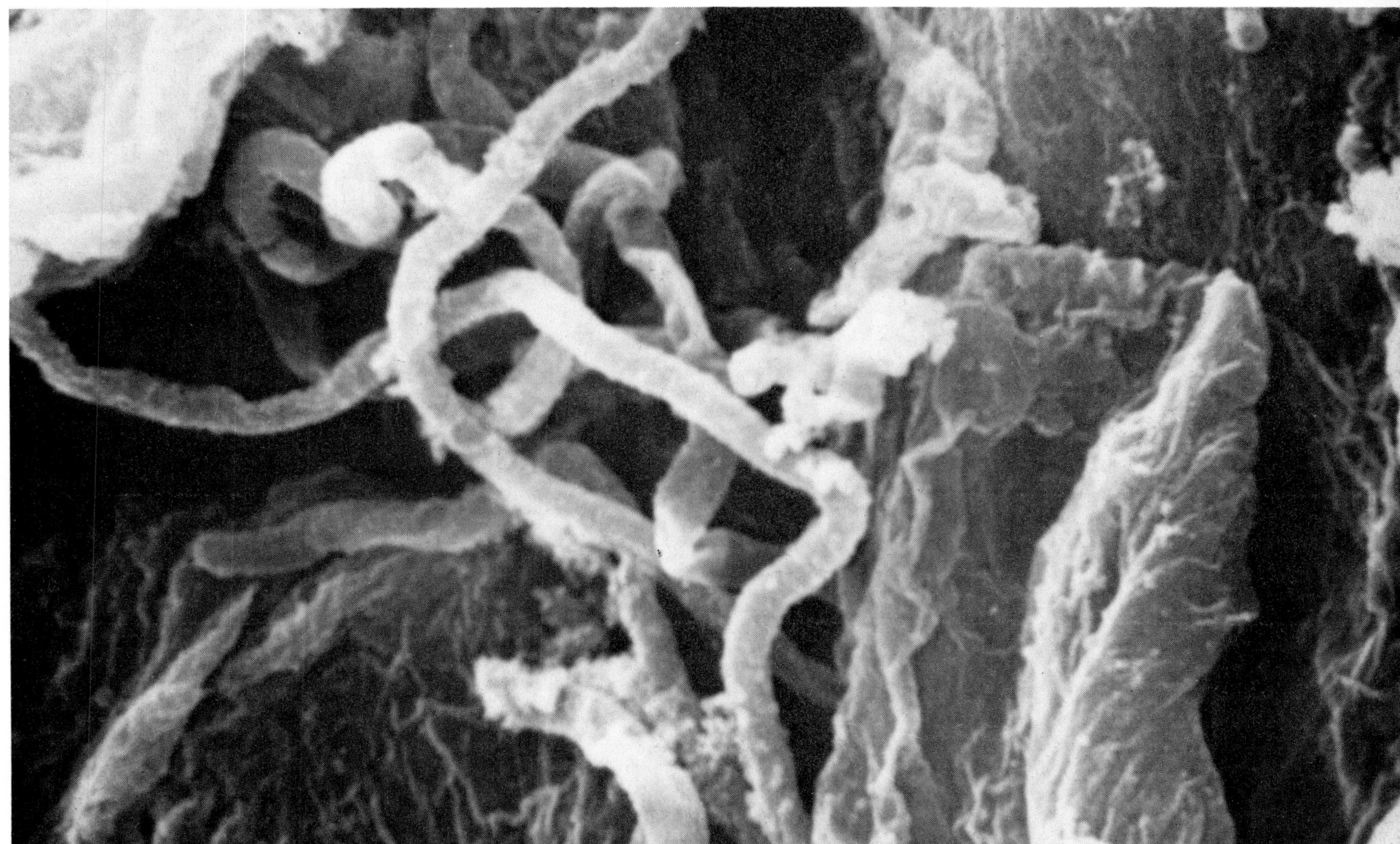

Streptococcus pyogenes, which causes scarlet fever, tonsillitis, and many other diseases, normally grows in long, gracefully looping chains. The individual healthy bacteria are almost perfectly spherical with smooth and featureless surfaces.
5200 X—Dr. D. Greenwood, St. Bartholomew's Hospital, London

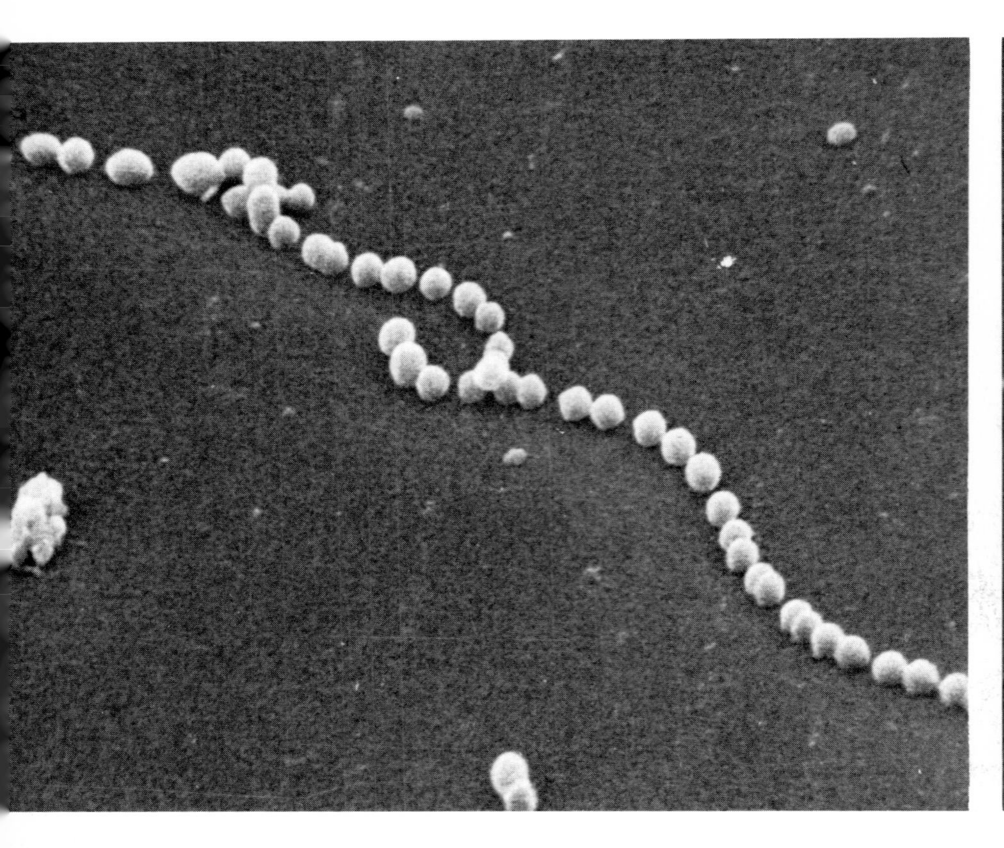

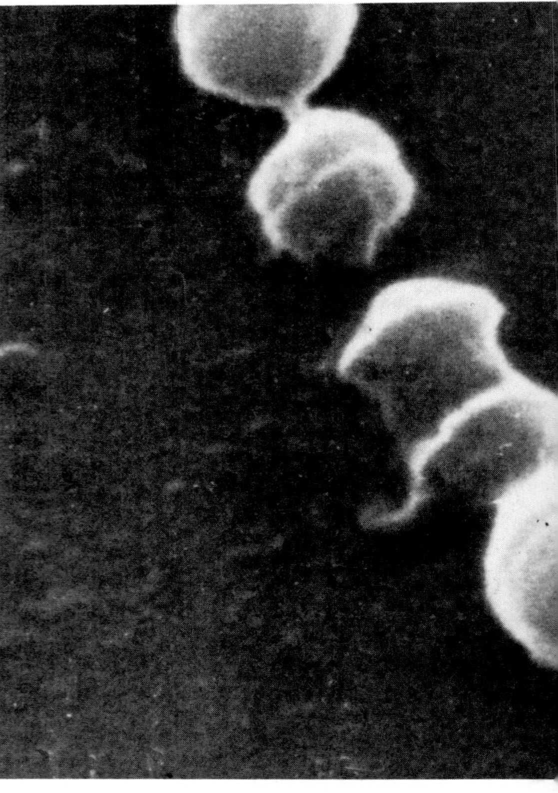

A totally different kind of reaction to the same antibiotic can be seen with *Staphylococcus aureus*, which causes boils, wound infections, and many other diseases. This organism is also spherical and smooth in a healthy condition, but grows in clusters instead of chains.
21,500 X—Dr. D. Greenwood, St. Bartholomew's Hospital, London

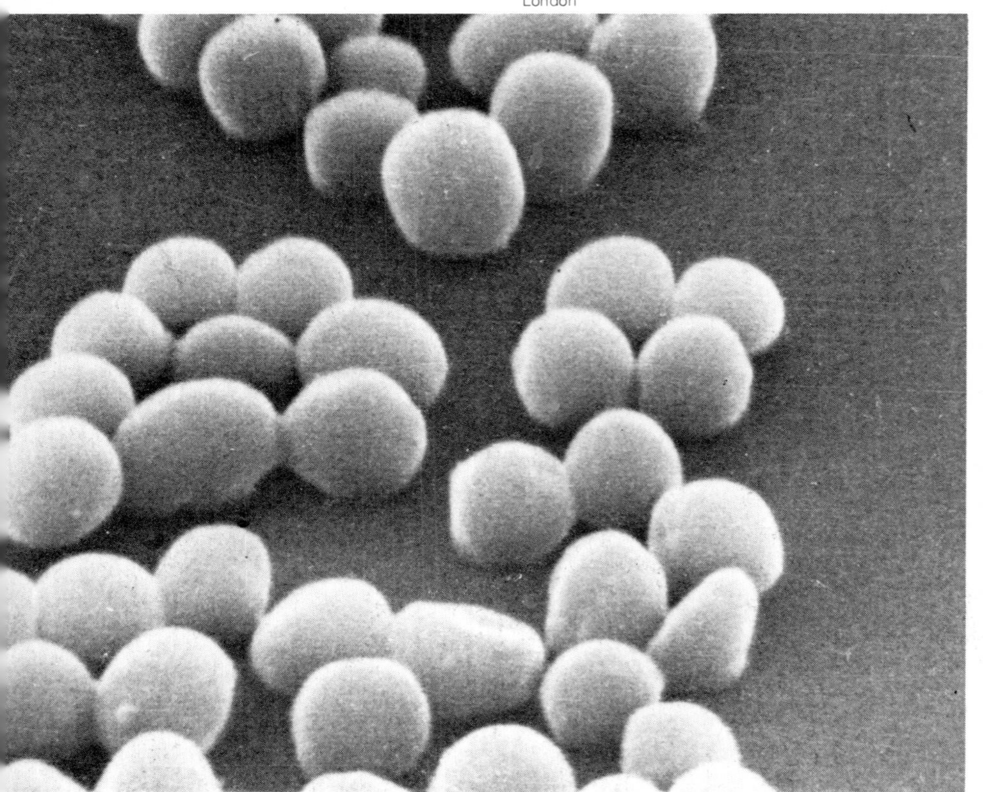

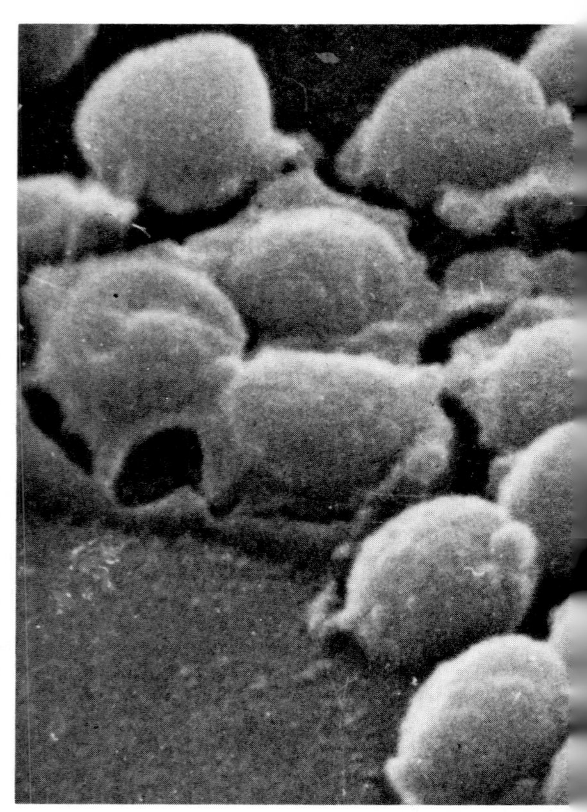

How Do Antibiotics Kill Bacteria?

These photographs give important clues.

With the scanning electron microscope, bacteriologists can see for the first time exactly how various doses of different drugs affect a wide variety of disease-causing bacteria. The technique thus not only yields new fundamental knowledge, but can help clinicians determine the proper course of treatment in human disease.

Exposed to a weak solution of synthetic penicillin, some individuals develop "apple-core" lesions, while other organisms show hardly any change. Bacteriologists had long thought that immature bacteria in the process of cell division are far more vulnerable to antibiotics than relatively mature bacteria. These photographs tend to confirm that hypothesis.
34,200 X—Dr. D. Greenwood, St. Bartholomew's Hospital, London

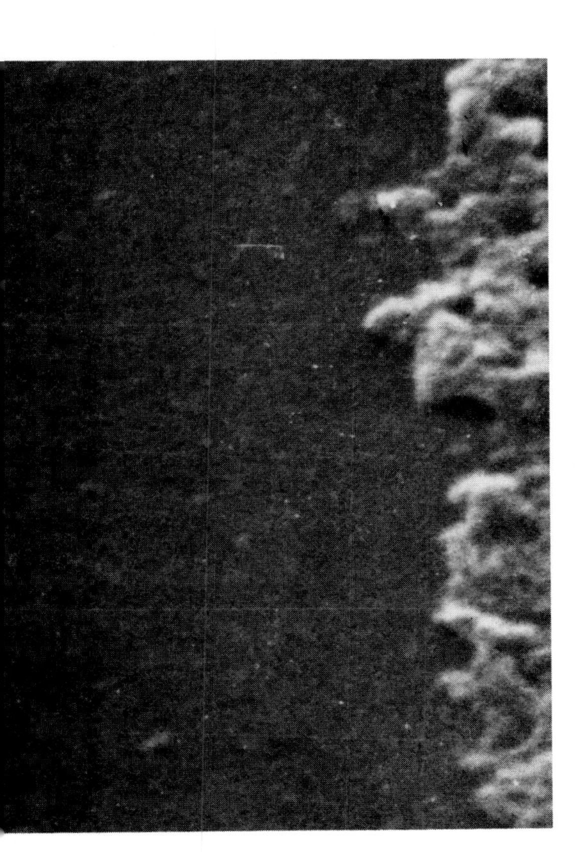

After exposure to penicillin, *Staphylococcus aureus* develops a rough, lumpy outer membrane—a sign that the penicillin is interfering with the constant process of cell-wall synthesis. The damage seems to occur almost everywhere rather than in a clearly defined ring as it did on *Streptococcus pyogenes*.
26,500 X—Dr. D. Greenwood, St. Bartholomew's Hospital, London

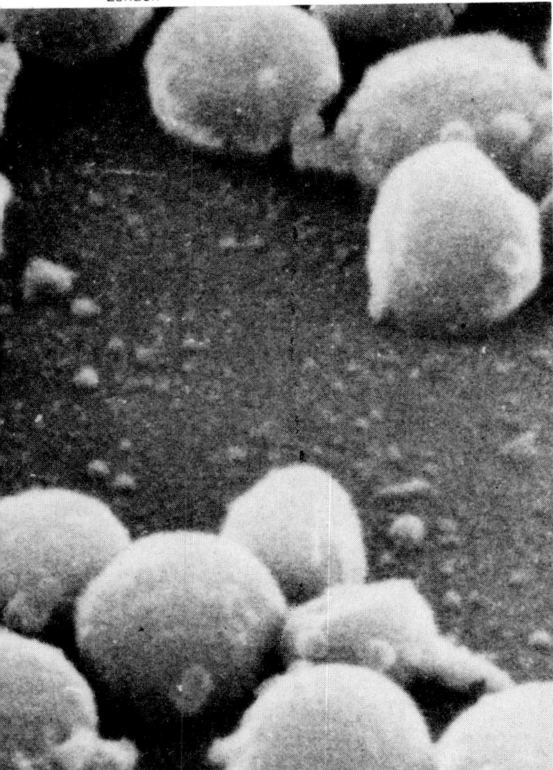

Same organism, different treatment. This culture of *Staphylococcus aureus* has been exposed to fusidic acid and individual organisms have collapsed like rubber balls with the air sucked out.
21,500 X—Dr. D. Greenwood, St. Bartholomew's Hospital, London

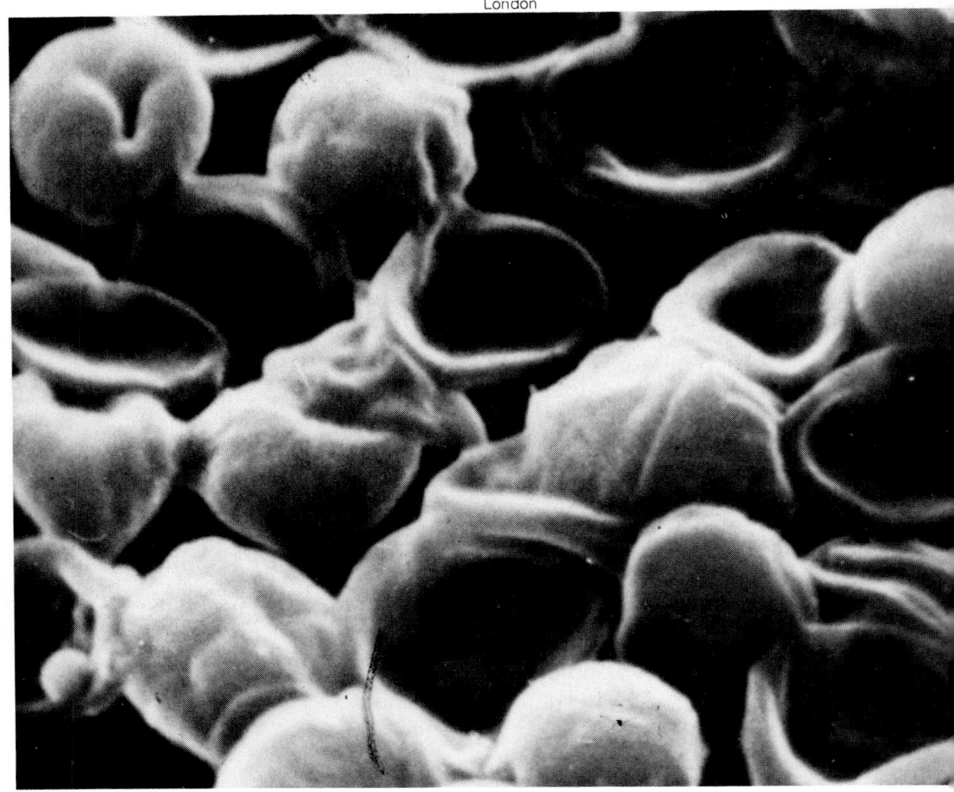

Effect of DDT on Bird Eggs

The brown pelican nests in only one place on the west coast of the United States: the starkly beautiful island of Anacapa, 10 miles off the coast north of Los Angeles. In the spring of 1971, only eight new babies were hatched, despite 540 nesting attempts. The year before it was one; in 1969, five. A normal season would produce 1000 to 1200 young birds. The apparent trouble was that most eggs had paper-thin shells. They were not strong enough to withstand the normal wear and tear of hatching, and were broken in the nests before the embryos could develop.

For some time it had been known that the pelican population—and the eggs—contained considerable quantitites of DDT. The original source is apparently runoff from sprayed fields and pesticide plant wastes. Eventually, the DDT makes its way to the ocean, where it is found in very dilute quantities. But the ocean's plankton absorbs some. The plankton are eaten by small surface fish such as anchovies, which in turn are eaten by the pelicans. Through this process of biological magnification, pelicans and other birds at the top of the food chain have been found with relatively high levels of DDT—as much as 1,800 parts per million in the birds themselves and in the broken egg shells in their nests.

Was the DDT causing eggshell thinning? There was much evidence to indicate that this was the case. Recently, a group at the University of California in Davis set up an experiment to measure the precise effect of DDT on birds' eggs. They fed Japanese quail a diet containing 5, 25, or 225 parts per million of DDT. The photographs show the results. Even low doses caused considerable degeneration of the eggshell, higher doses produced dramatic results. The photo near right is a normal, healthy quail egg shell. Opposite is the shell of an egg laid by a quail fed 225 parts per million DDT for 57 days. The shell is 25 percent thinner, with 60 percent of the spongy layer gone. The spongy layer is the thick central area that gives the shell most of its strength. The whole shell shows extreme cavitation, or development of holes. Shells of broken eggs from pelicans' nests show similar defects.

1150 X (p. 130); 1250 X (p. 131)—L. Z. McFarland, R. L. Garrett, J. A. Nowell, University of California, Davis

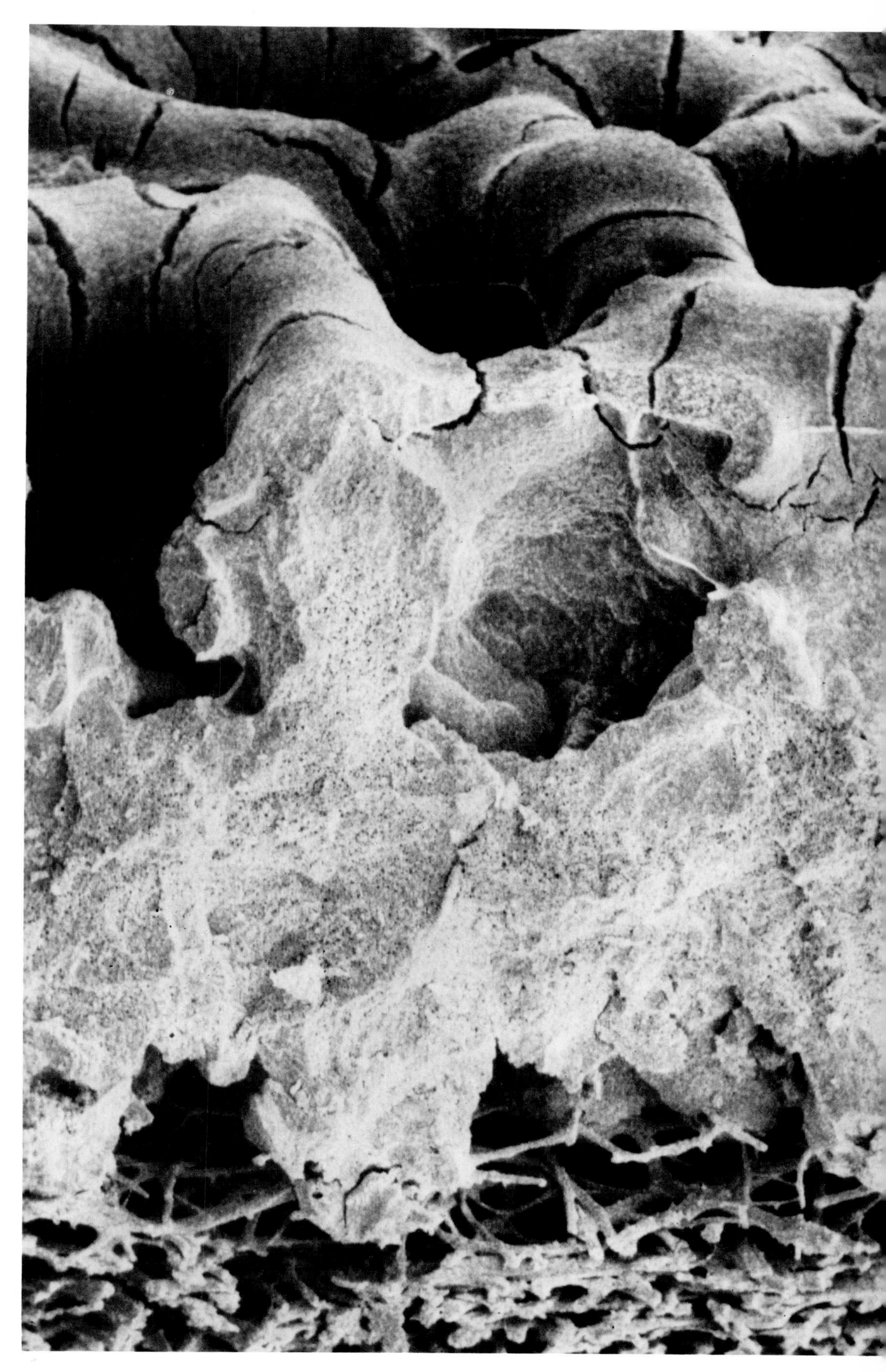

Slime Mold

After feeding, the amoebae gather together and develop a thin, sheath-like canopy that covers groups of cells. ▶
17,000 X—Dr. Robert P. George, University of Wisconsin

The lower orders of the animal and vegetable kingdoms sometimes are so similar that it is difficult to tell to which division they belong; indeed, they display properties of both. Undoubtedly one of the most fascinating of these ambiguous life forms is a class that goes by the unlovely name of slime mold. The general pattern of behavior of slime molds has been known for many years; a biologist named Kenneth Raper isolated the species known as *Dictyostelium discoideum* (shown here) from a batch of decaying leaves he found on the forest floor in 1935. Yet not until the scanning electron microscope was it possible to make such a stunning photographic record of the astonishing behavior of this plant/animal.

When the amoebae of the slime mold *Dictyostelium discoideum* begin to grow, they feed upon bacteria and gather into formless blobs such as these resting on the surface of a laboratory filter.
8300 X—Dr. Robert P. George, University of Wisconsin

A larger-scale view of the surface of the filter shows that many cells have begun to form aggregation centers, again covered by the membrane. ▶
500 X—Dr. Robert P. George, University of Wisconsin

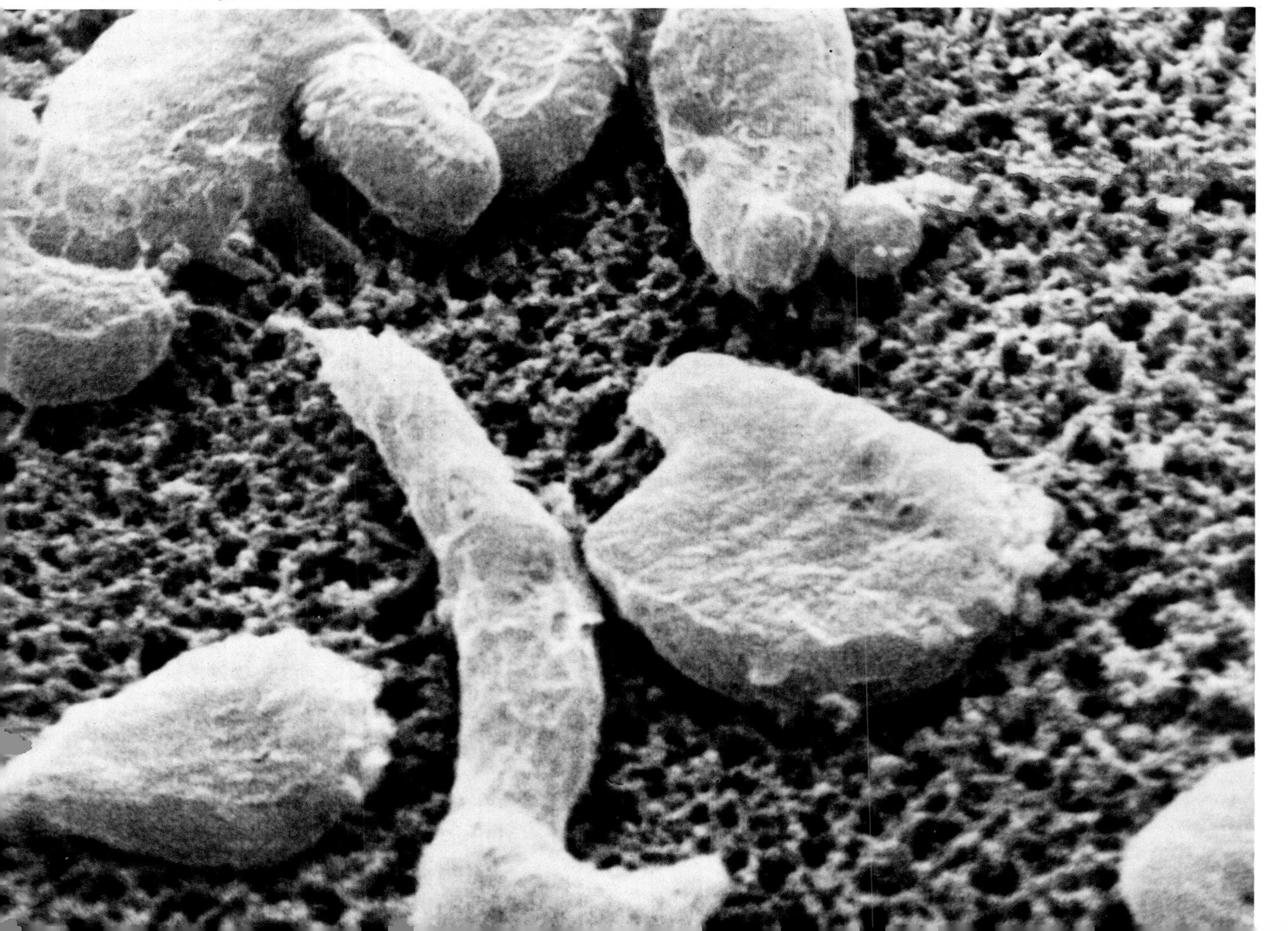

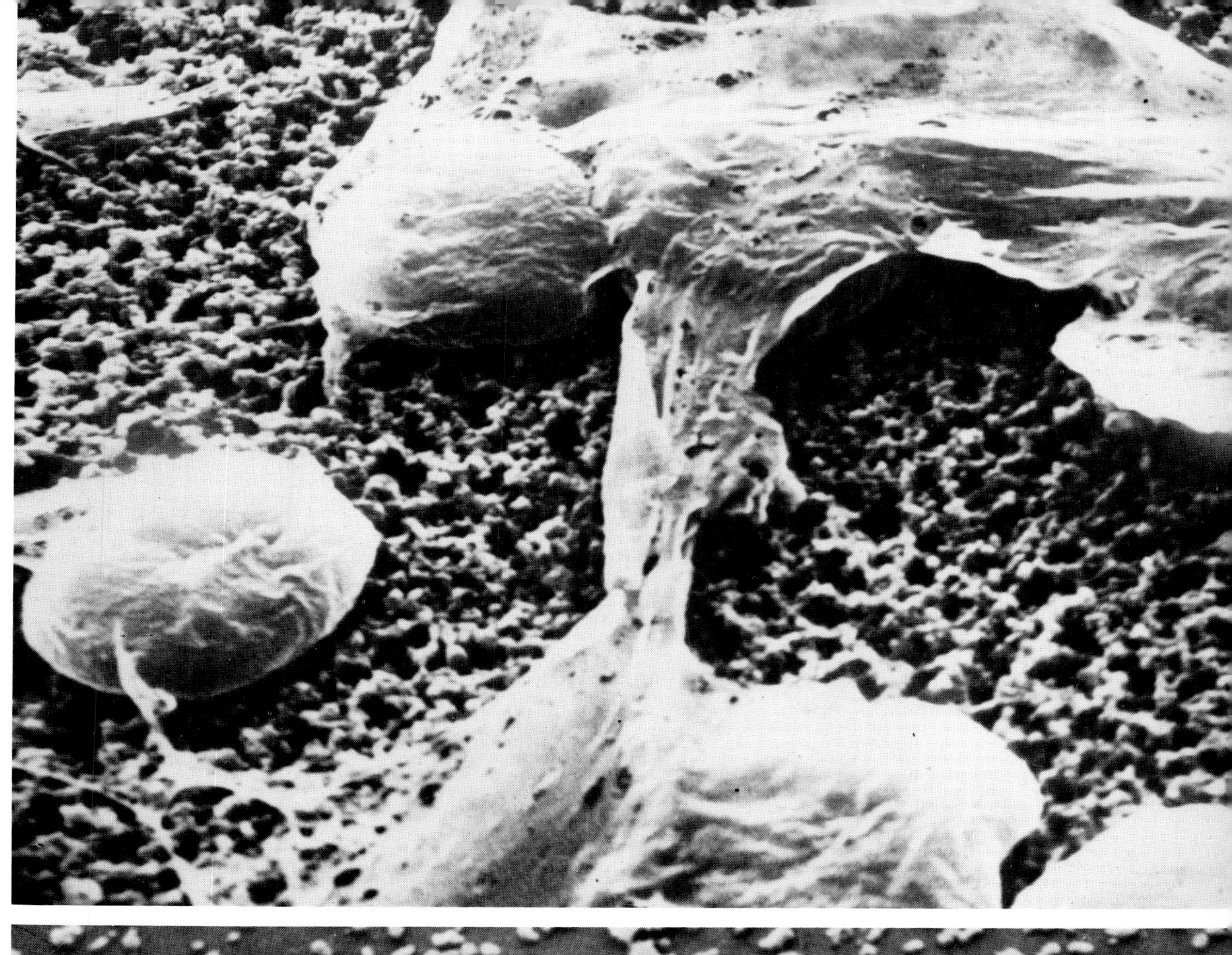

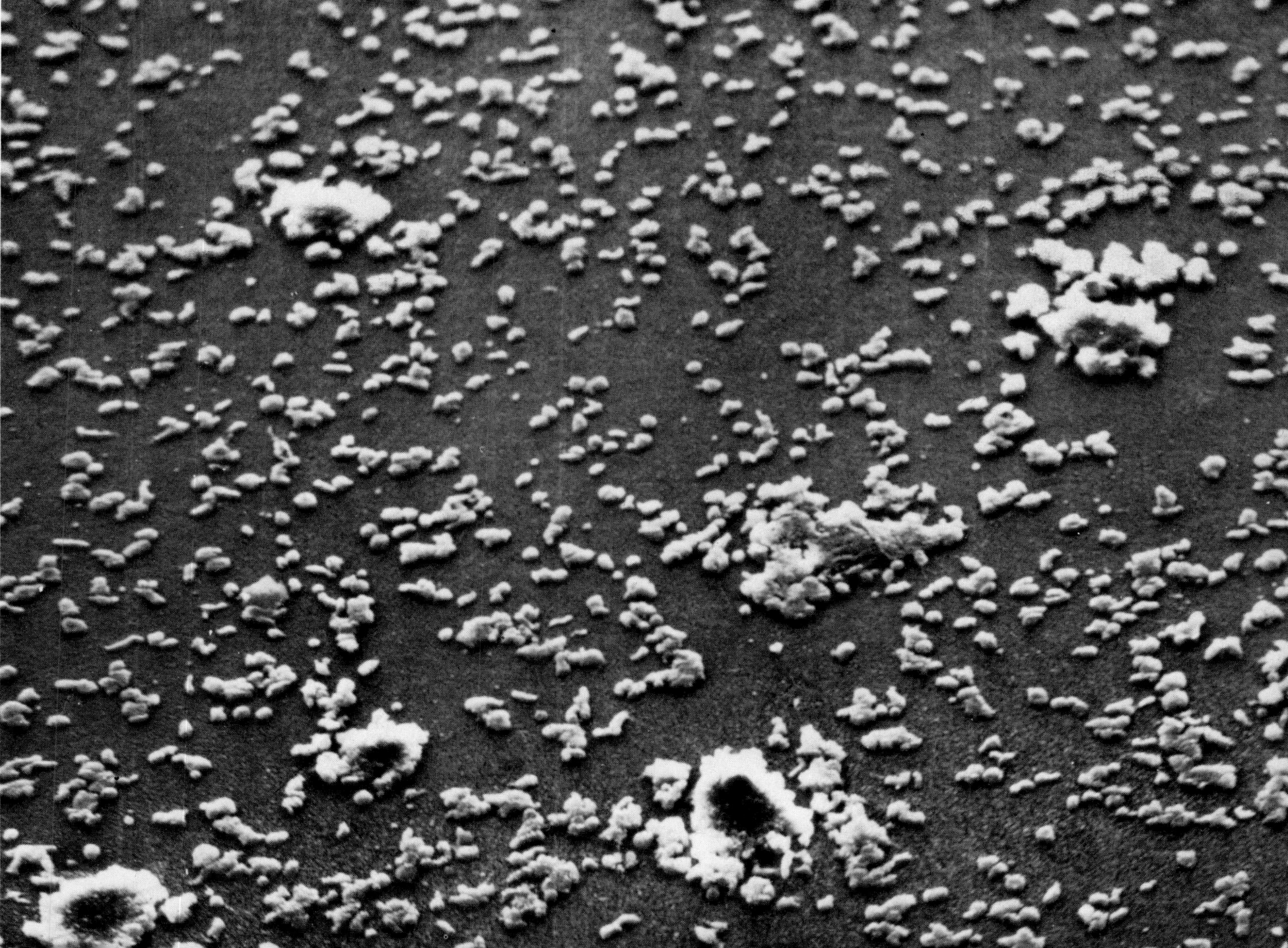

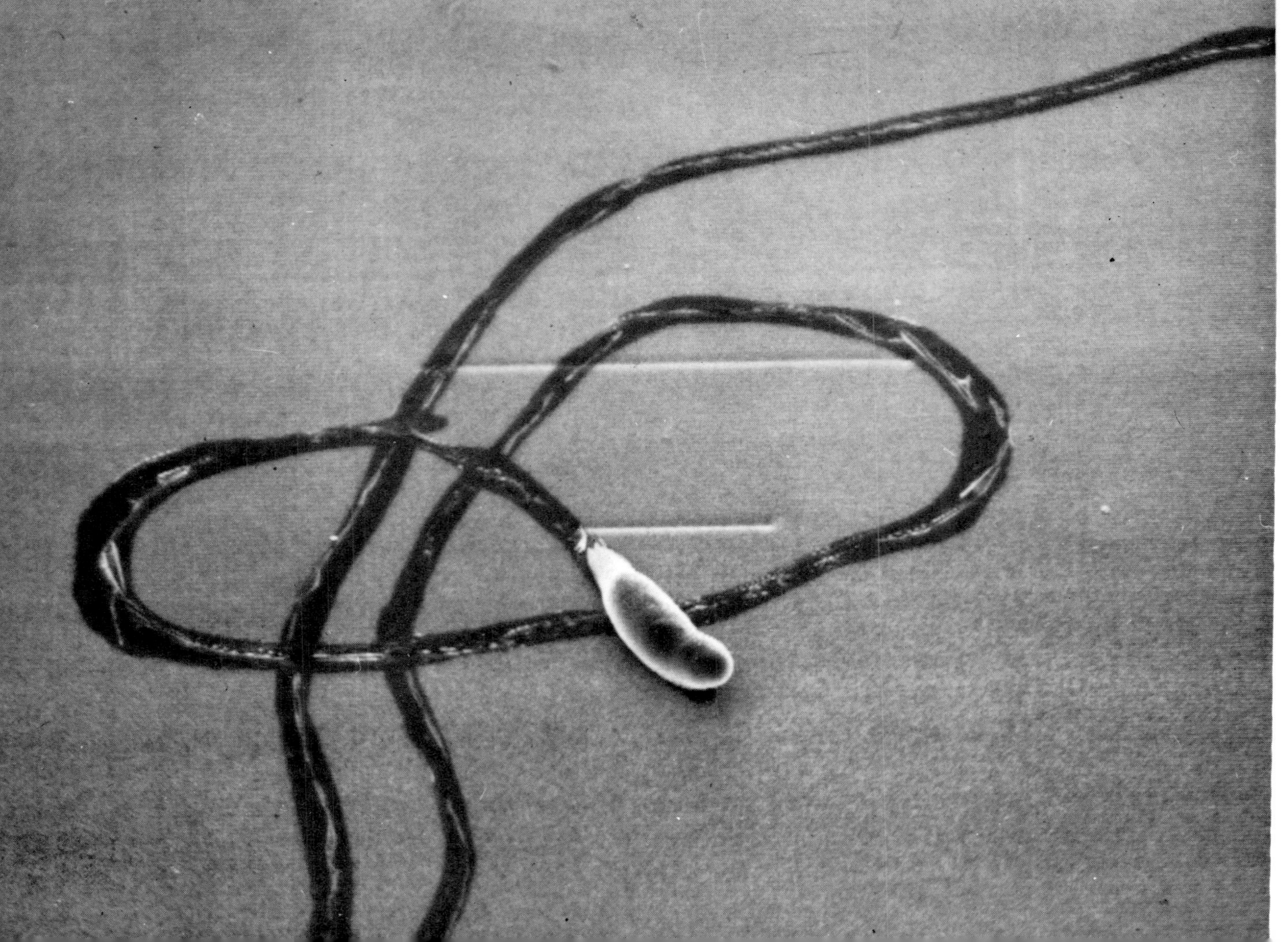

◄ **A slug-like structure forms at the middle of each aggregation of cells. As it reaches its full height, it bends over until the tip touches the surface on which the cells are resting.**
3400 X—Dr. Robert P. George, University of Wisconsin

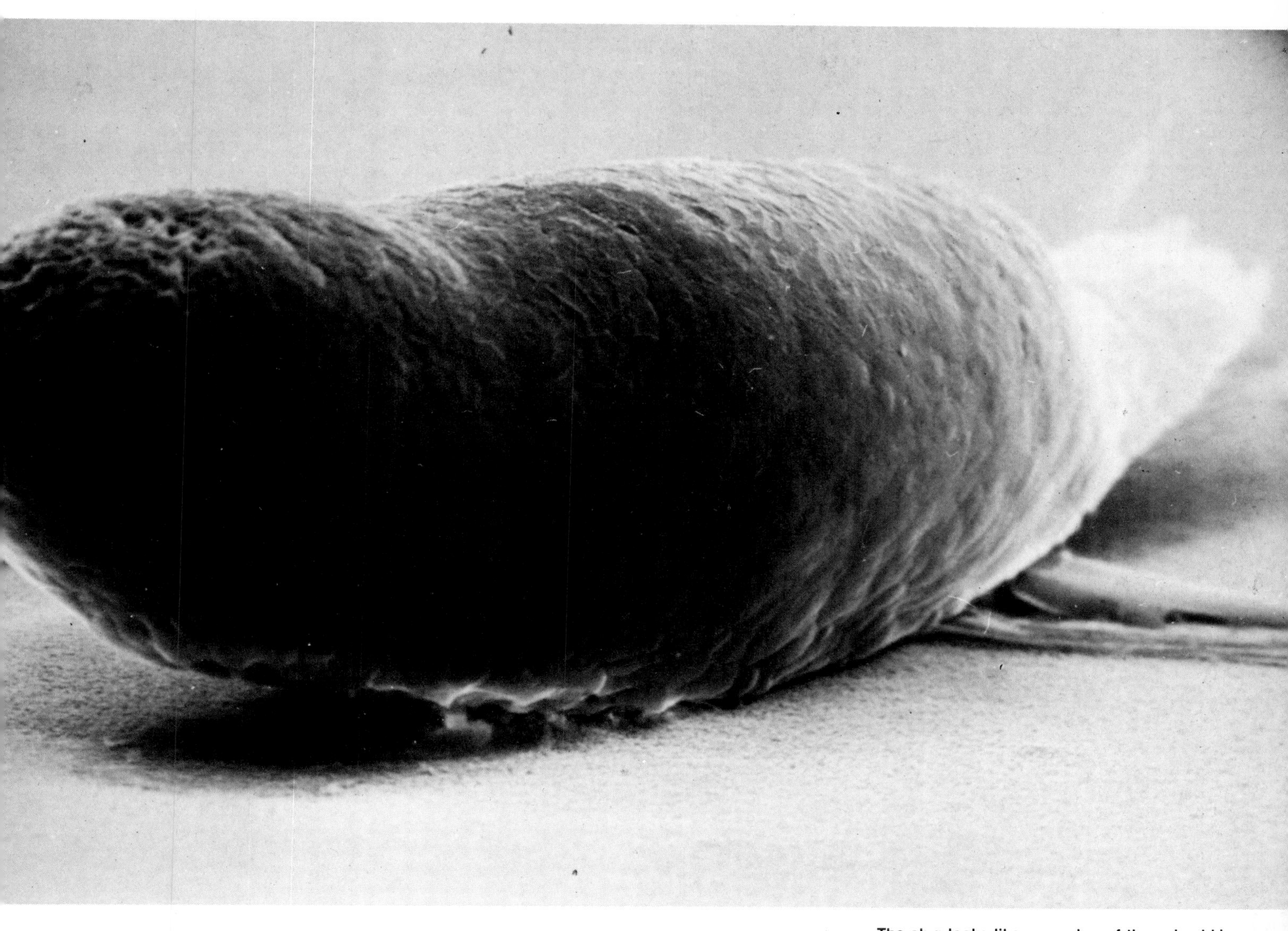

The slug looks like a member of the animal kingdom. But it is not a single creature; it is a collection of individual and separate cells that have organized themselves into a cooperative enterprise.
1200 X—Dr. Robert P. George, University of Wisconsin

◄ **Then it breaks loose from the base on which it formed and slowly begins to creep forward, leaving a trail of slime from which it gets its name.**
130 X—Dr. Robert P. George, University of Wisconsin

After a period of migration, the slug stops, forms a base, and grows into a plant-like structure with a pod at the end of a long, slender stalk. The slime trails made by the slug moving to the point where it takes root can be clearly seen. One slug, behind the stalk, is still looking for a place to stop. Another to the right has stopped, and is moving into an upright position to begin stalk formation.
130 X—Dr. Robert P. George, University of Wisconsin

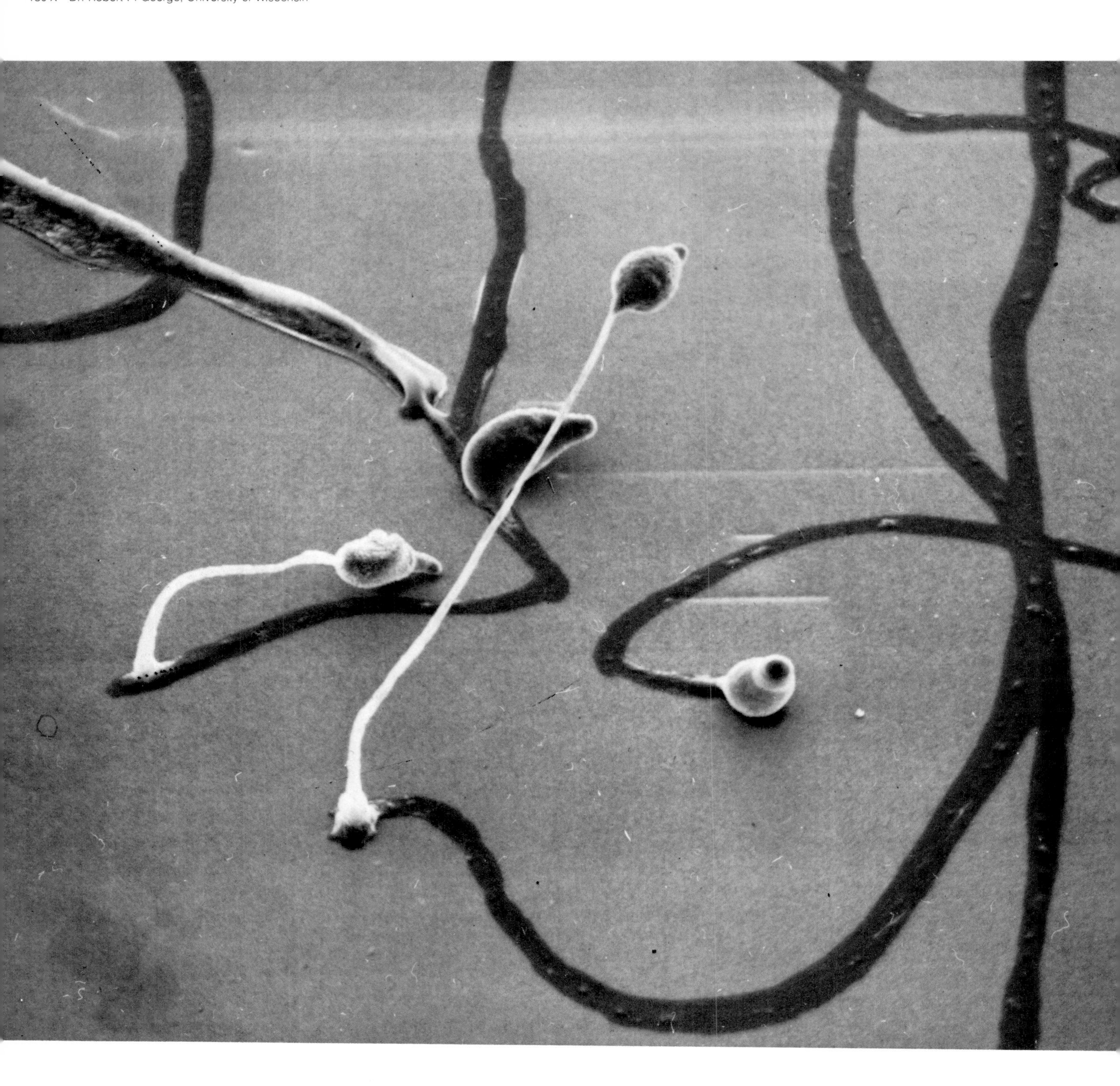

A closeup of the head, or "fruiting body," shows individual spores in place. Soon they drop away and begin the whole process again.
2450 X—Dr. Robert P. George, University of Wisconsin

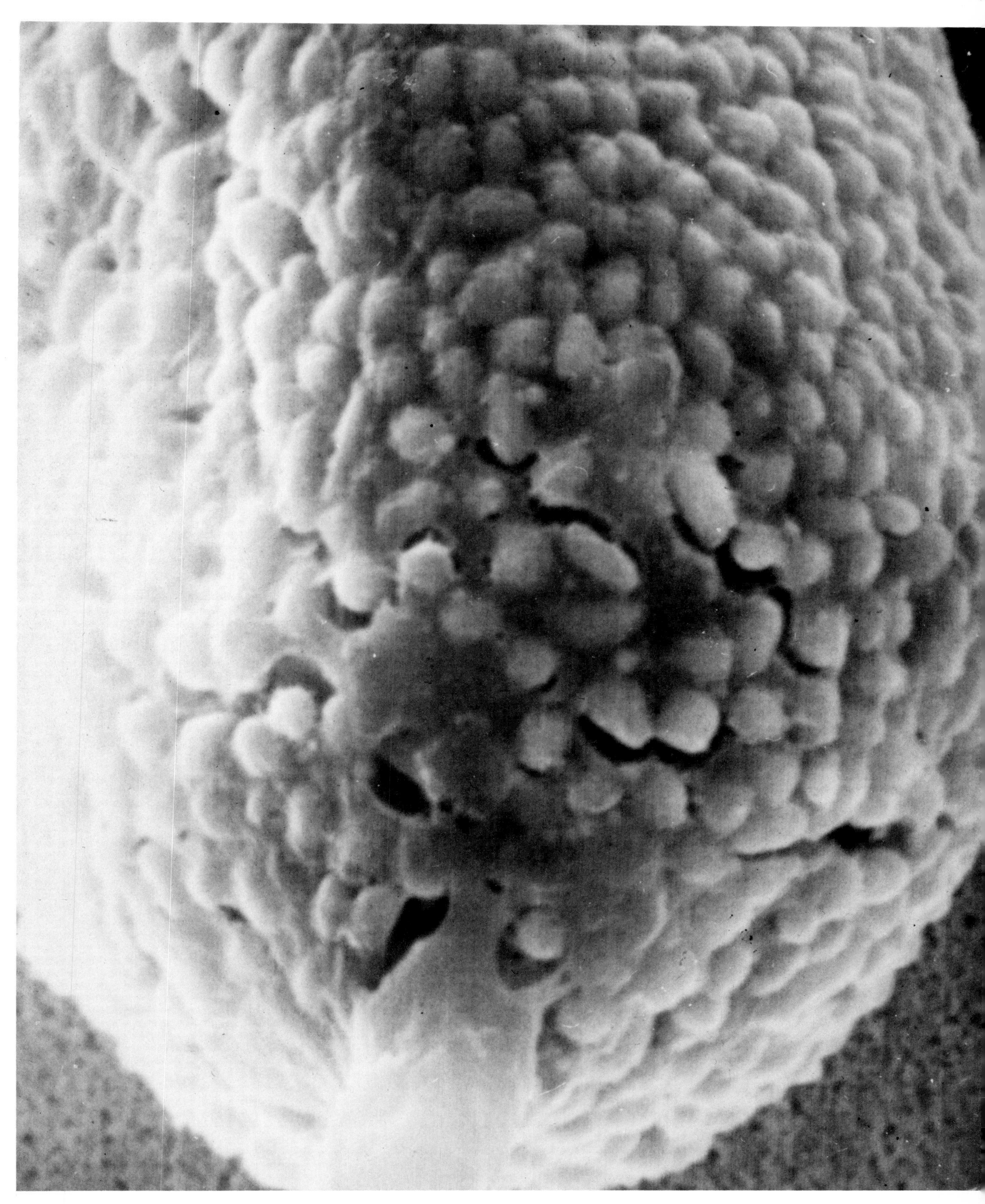

Pollen

Pollen grains carry the male genetic material of flowering plants. Thus these exceedingly small bits of intricately structured material are analogous to sperm in animals. Because the pollen from each plant has its own characteristic shape, it is an important research tool. Pollen grains preserved in bogs for hundreds of thousands of years, and even older fossilized pollen, can tell researchers what kinds of plants grew at certain periods, thus revealing the history of climate through the ages.

Pollen grows and matures in a part of the flower called the anther. It eventually travels to the female organ, the pistil, of the same flower or of another flower of the same species, where it is trapped by a sticky substance. If the fertilization is successful, the pollen puts out an extension that penetrates the surface and grows downward between the cells of the pistil until it reaches the ovary. There, a nucleus from the pollen fuses with an egg nucleus to form an embryo, from which grows the fruit of the plant.

Most pollen is extremely small. The largest grains are about 250 microns (a quarter of a millimeter) in diameter and are just visible to the naked eye. Most are much smaller, with the smallest perhaps two microns across.

Some pollen is carried from one plant to another by the wind, some by insects; other pollen grains fall on and fertilize the same flower that produced them. Plants pollinated by insects are usually colorful and have a conspicuous odor; sometimes they have nectar that attracts certain insects, and a structure designed to accommodate precisely the operative insect but almost no other. The flowers of wind-pollinated plants are usually much less showy and do not produce nectar, but they do release huge amounts of pollen, since much of it will obviously miss its target when wafted randomly by the wind. Some evergreen trees produce so much pollen that it can be seen as a cloud over the forest in windy weather.

The structure of the pollen grain has been studied for years, but never with so much success as with the scanning electron microscope, which, as these pages show, has made it possible to see the intricate and often lovely structure of these fascinating grains in superb detail.

The outer surface of pollen is composed primarily of an extremely tough substance called sporollenin. It resists almost everything—even many acids—and is frequently preserved for thousands of years.

The structures of different pollen grains, as these spectacular pictures show, are incredibly rich and varied. Some are clusters of spines, or bacula. Some look like surrealist lunar landscapes, others like fruits, yet others are covered with baroque-patterned balustrades. As might be expected, the larger grains, which are usually carried by insects rather than the wind, tend to have more elaborate structure. A number of investigators in laboratories around the world are now using the scanning electron microscope to investigate pollen in the attempt to learn why it assumes the fantastic variety of shapes and patterns seen on these pages. In the process they are discovering in yet another area the apparently limitless inventiveness of nature.

The reticulately patterned wall of a lily pollen grain ► (Lilium longiflorum). Some of the normal amorphous material has been removed by a chemical process to reveal the intricate structure of the raised walls, or muri. These balustradelike features are set on cylindrical columns, the baculae, which compare with the supports of the balustrade. The surface of another pollen grain is seen in the background.

9500 X—Professor J. Heslop-Harrison, Dr. Y. Heslop-Harrison, Royal Botanic Gardens, Kew, England

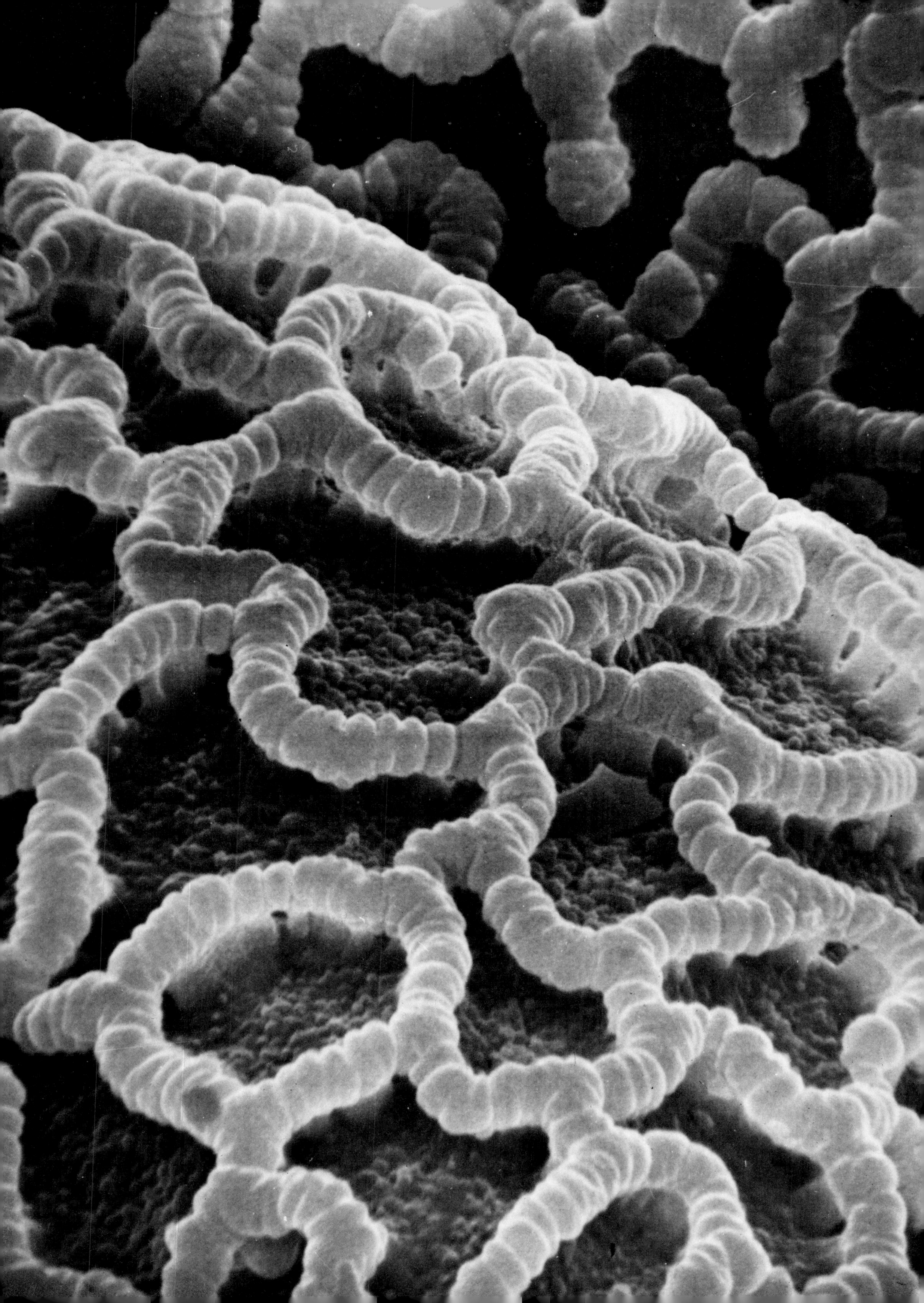

Lily pollen at the time of the release of the grains from the anther.
1850 X—Professor J. Heslop-Harrison, Dr. Y. Heslop-Harrison, Royal Botanic Gardens, Kew, England

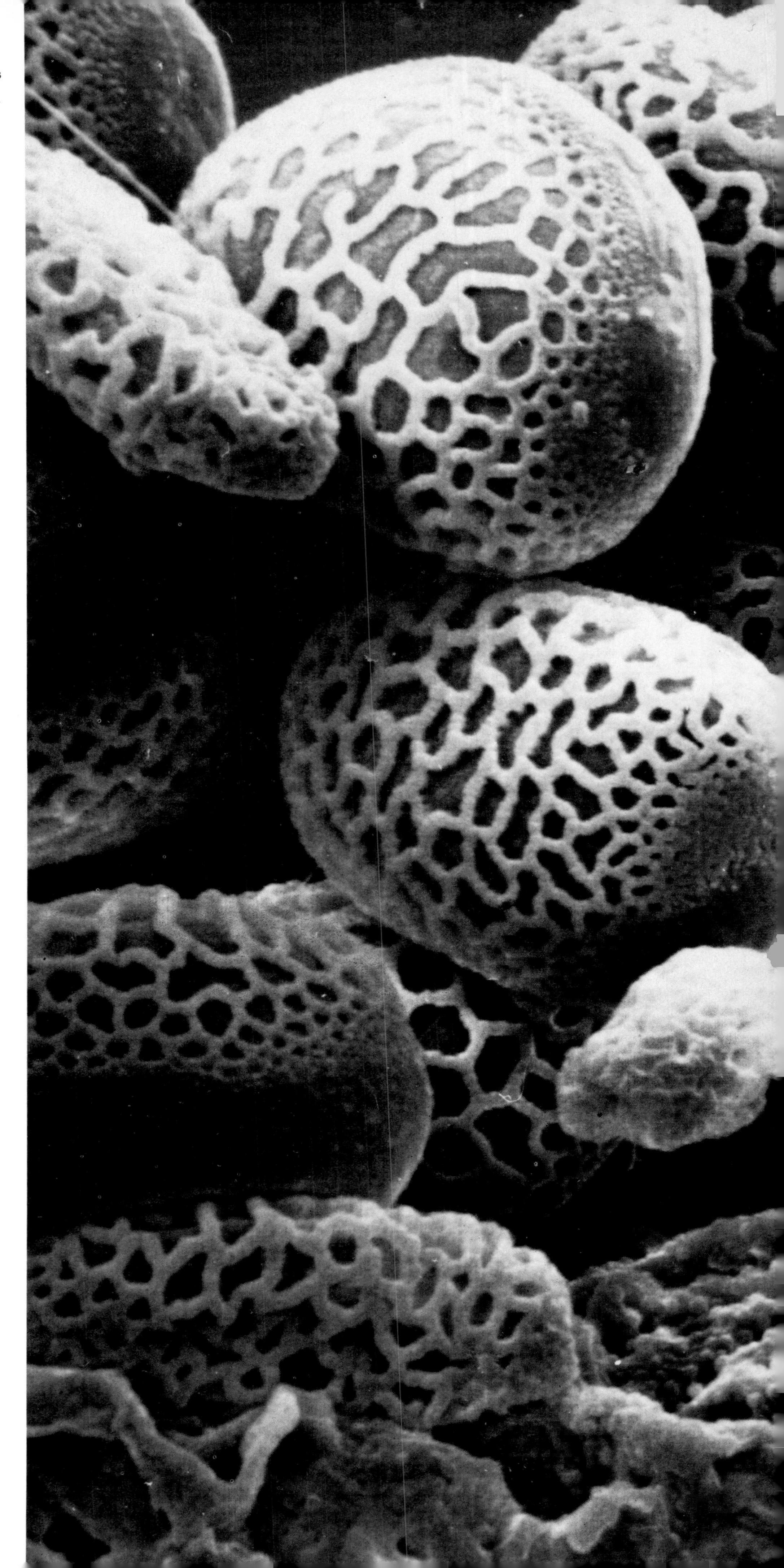

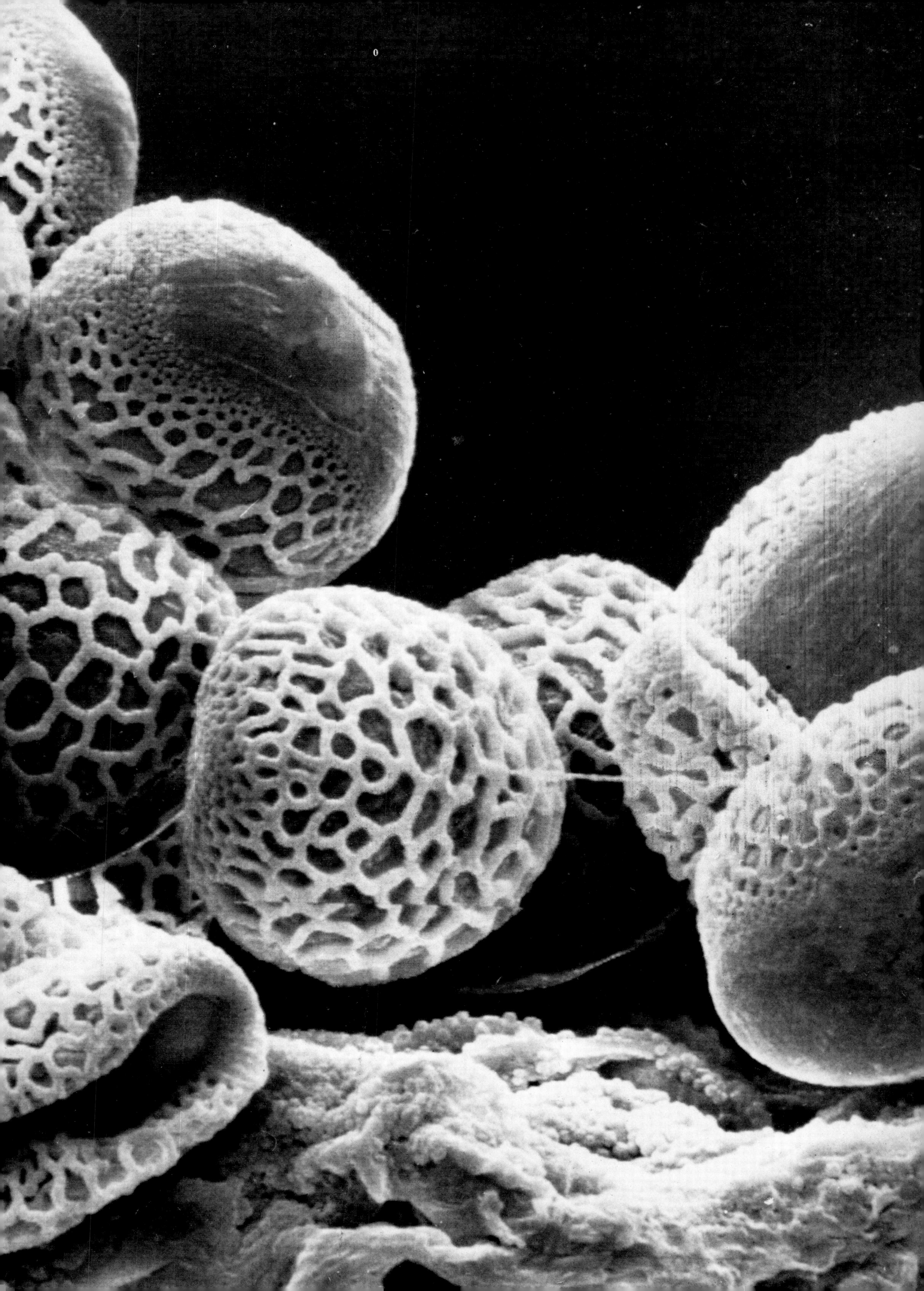

Pollen grain from *Cosmos bipinnatus*
6500 X—Professor J. Heslop-Harrison, Dr. Y. Heslop-Harrison, Royal Botanic Gardens, Kew, England

Beech-tree pollen ▶
2400 X—Miss B. Farber, Johnson and Johnson Research

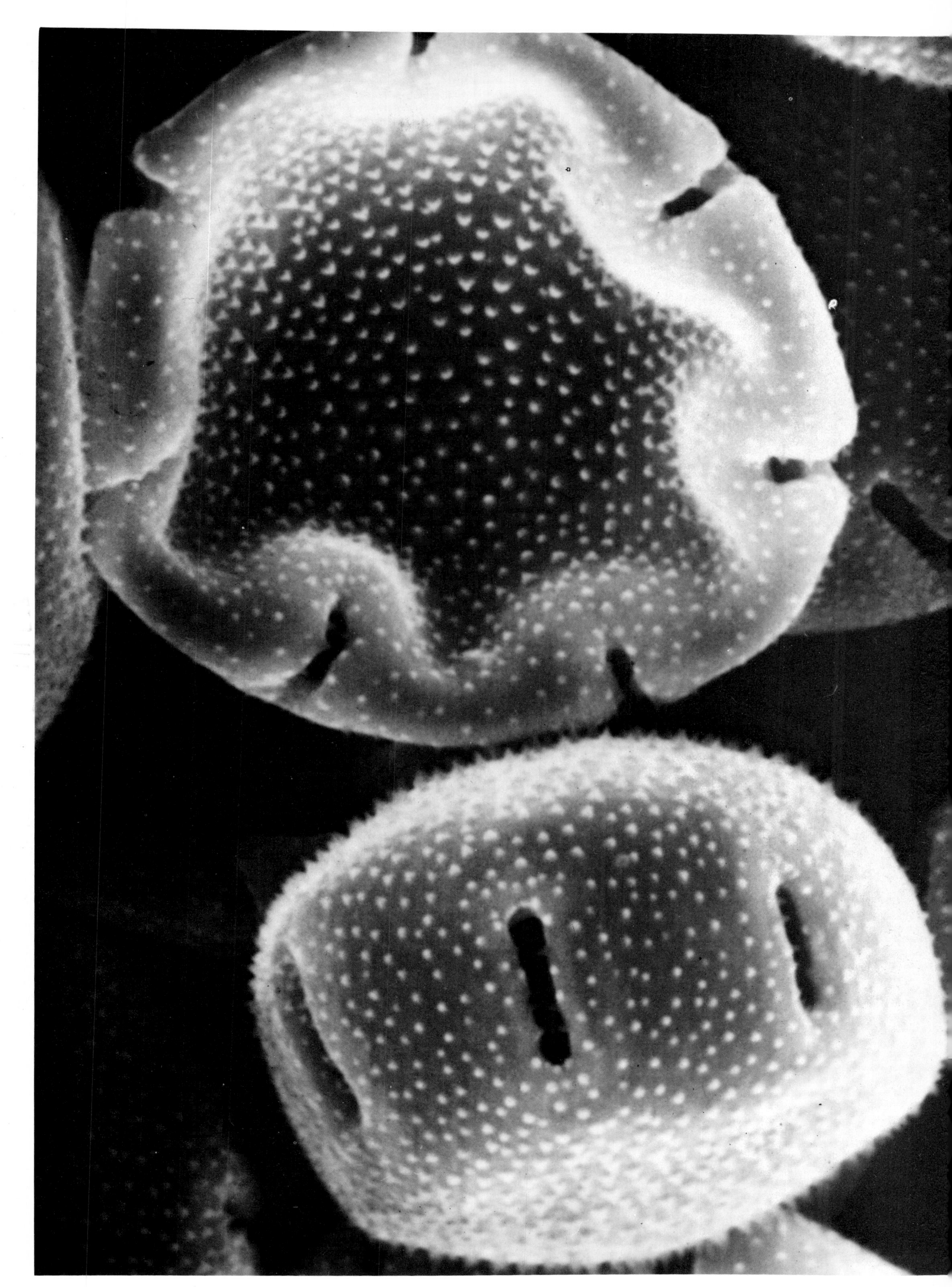

Pine pollen
3650 X—Dr. L. M. Beidler, Florida State University

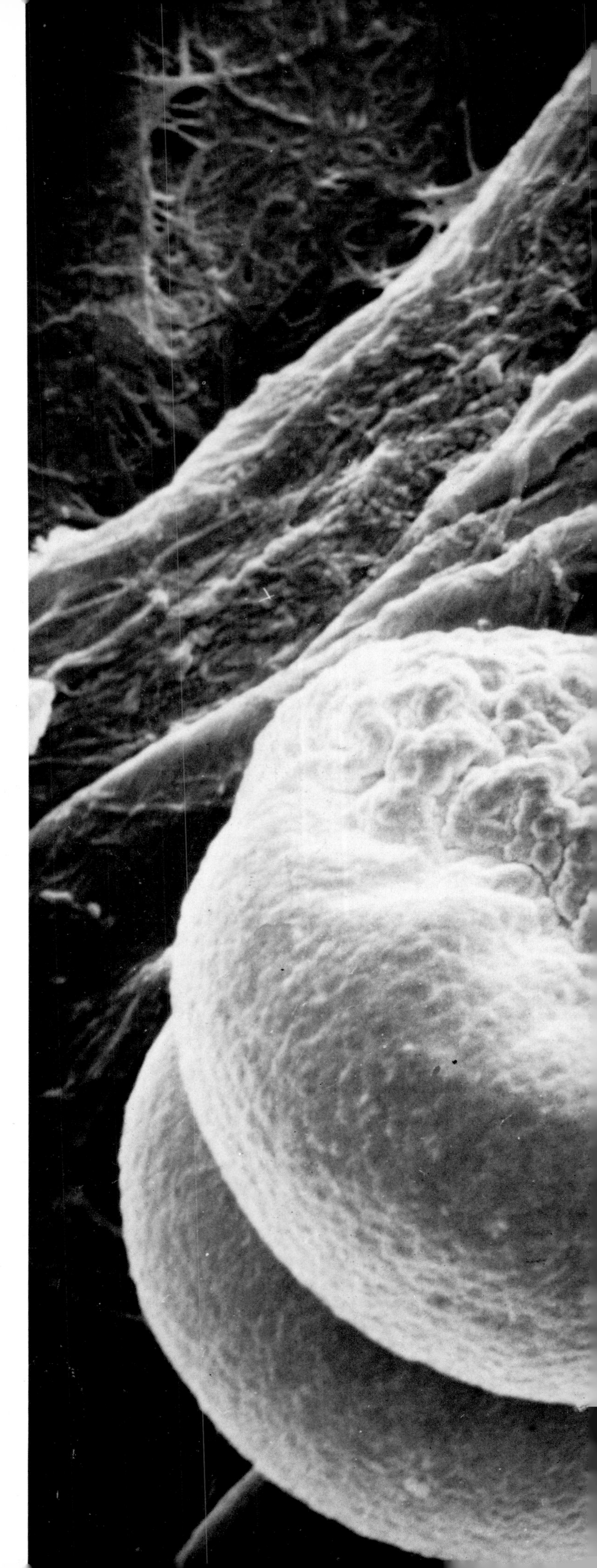

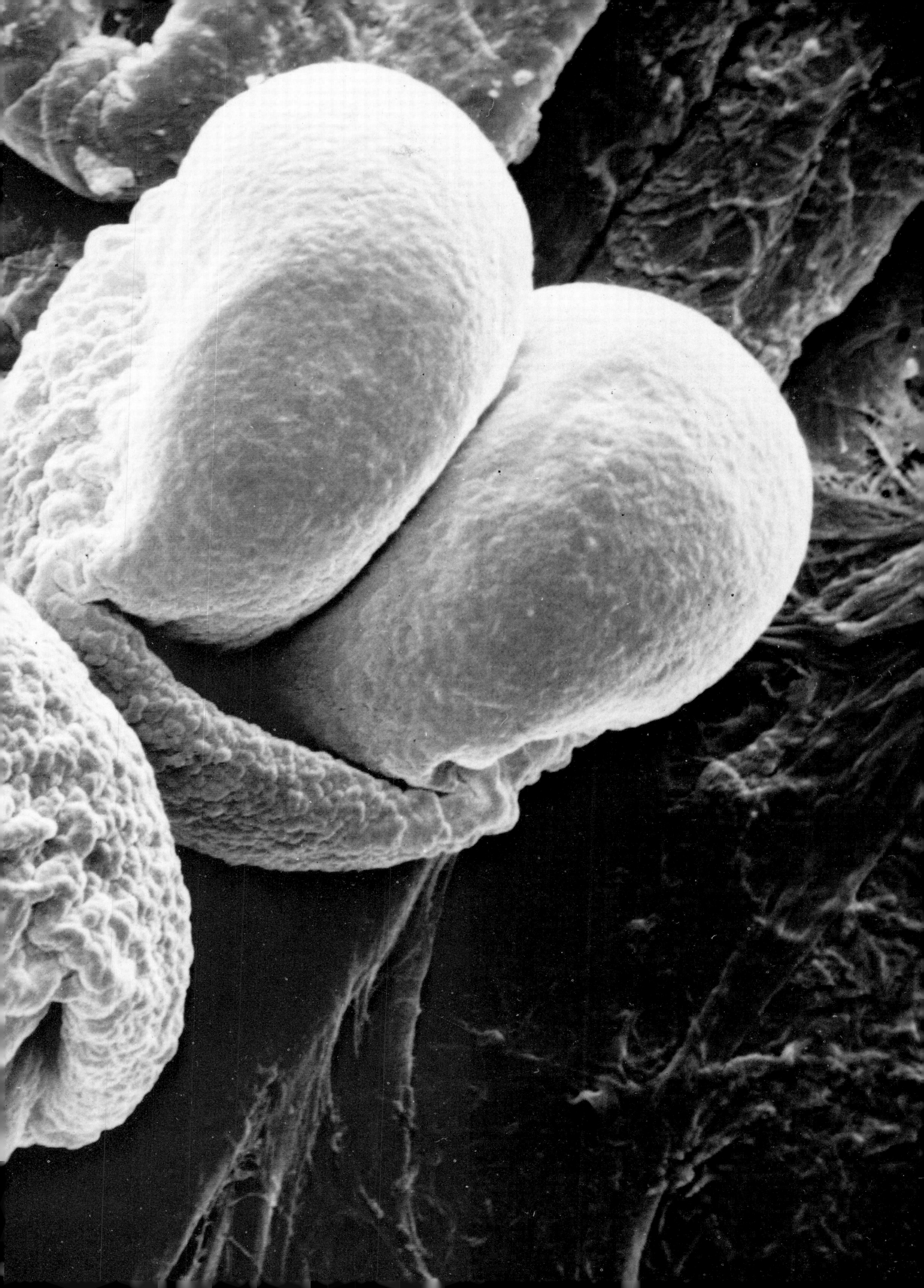

Ragweed pollen
5000 X—Richard Turnage, Advanced Metals Research Corporation

Hellebore pollen
5750 X—Dr. Patrick Echlin, University of Cambridge, England

next page: Morning glory pollen ▶
2500 X—Dr. Patrick Echlin, University of Cambridge, England

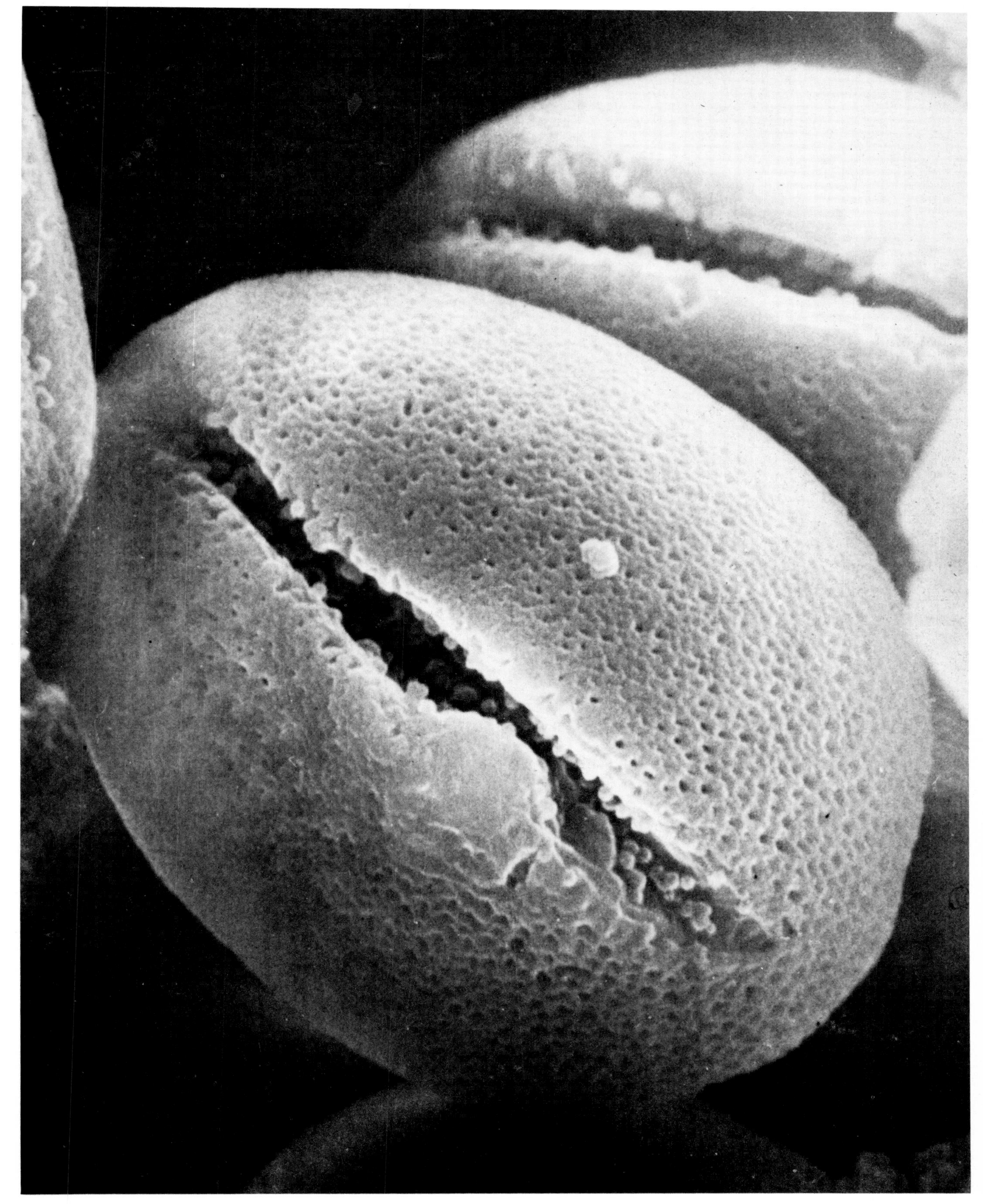

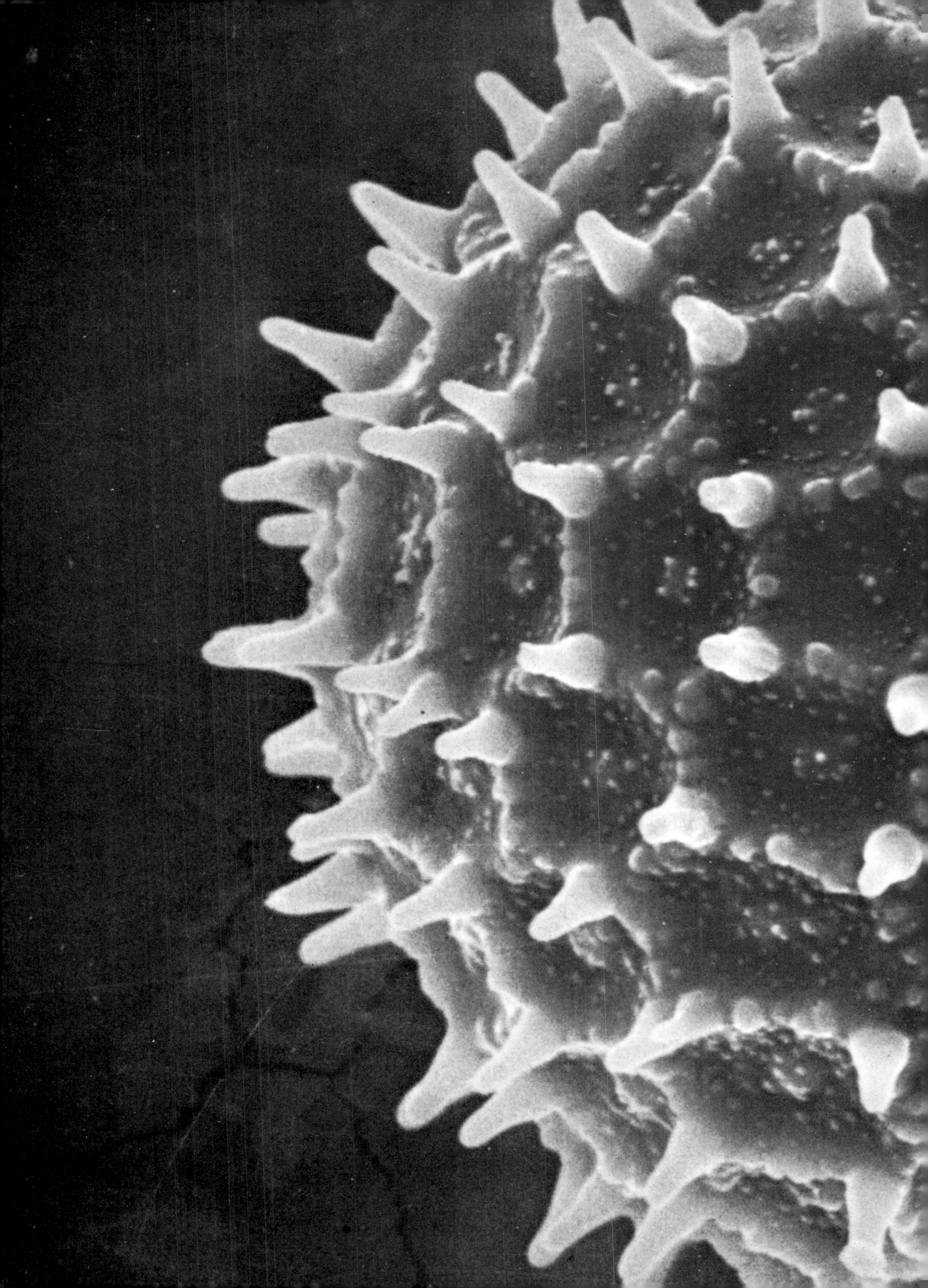

Thrift pollen
1080 X—Dr. Patrick Echlin, University of Cambridge, England

Flax pollen
3500 X—Dr. Patrick Echlin, University of Cambridge, England

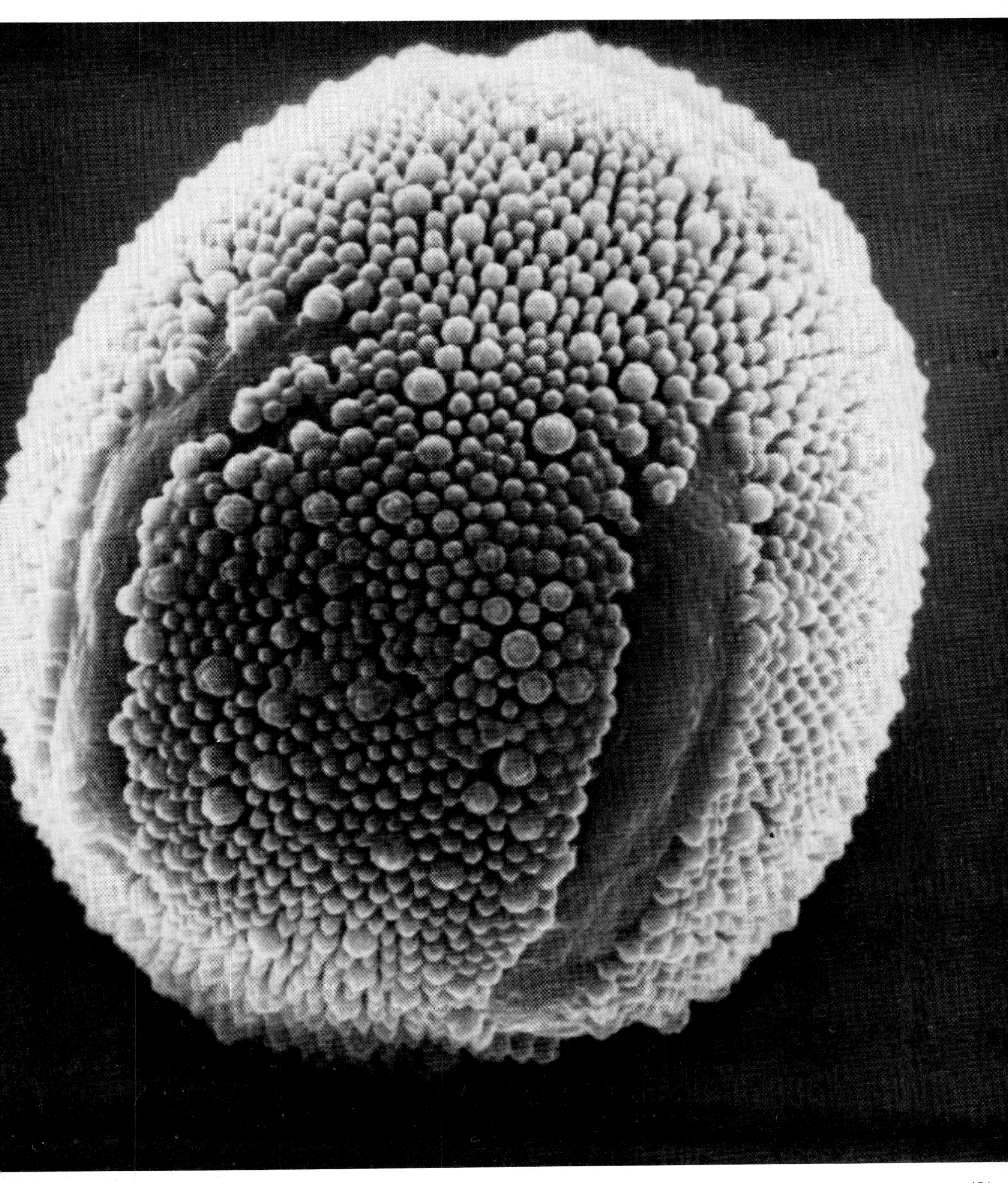

Wallflower pollen
2100 X—Dr. Patrick Echlin, University of Cambridge, England

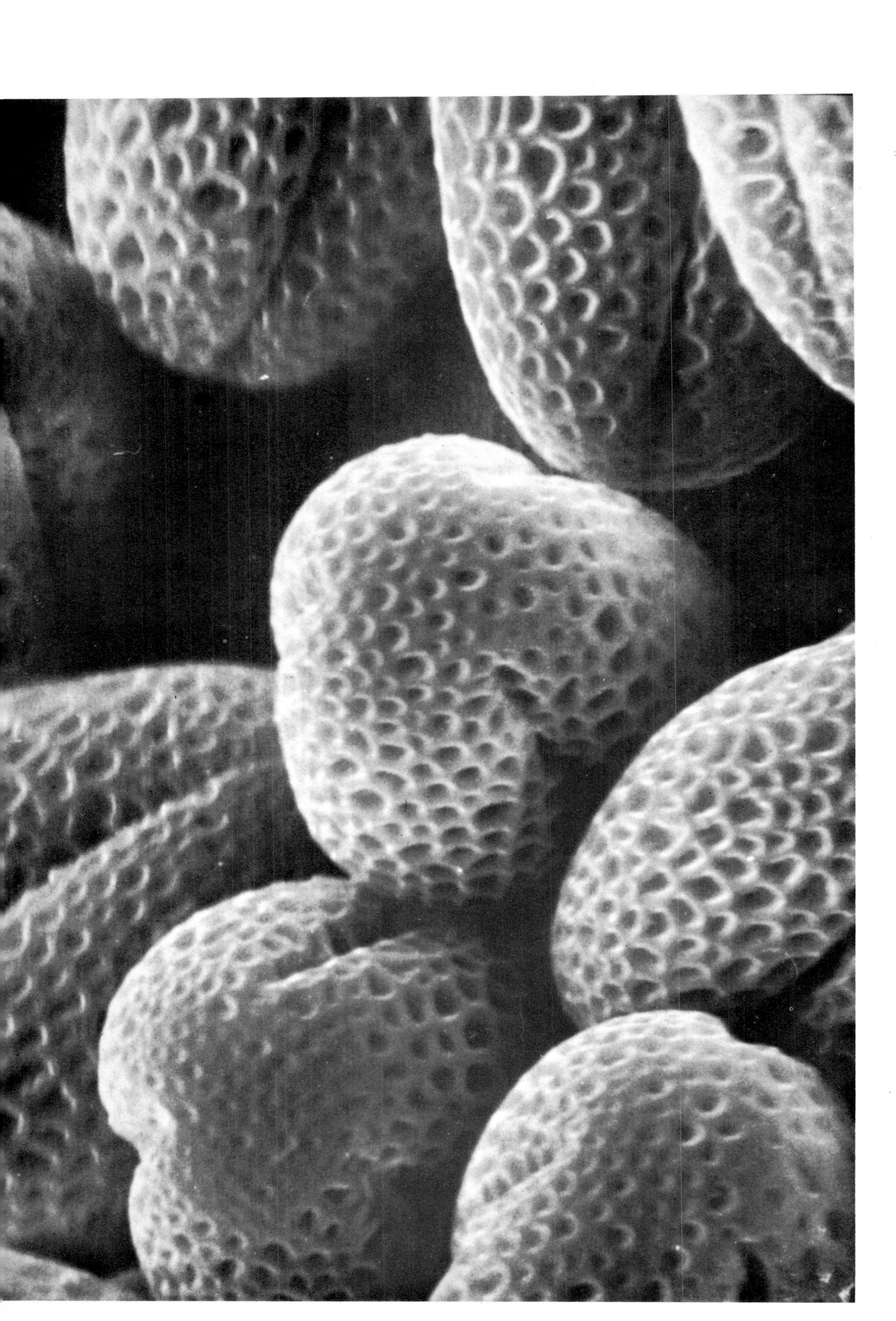

Touch-me-not pollen
13,250 X—Dr. Patrick Echlin, University of Cambridge, England

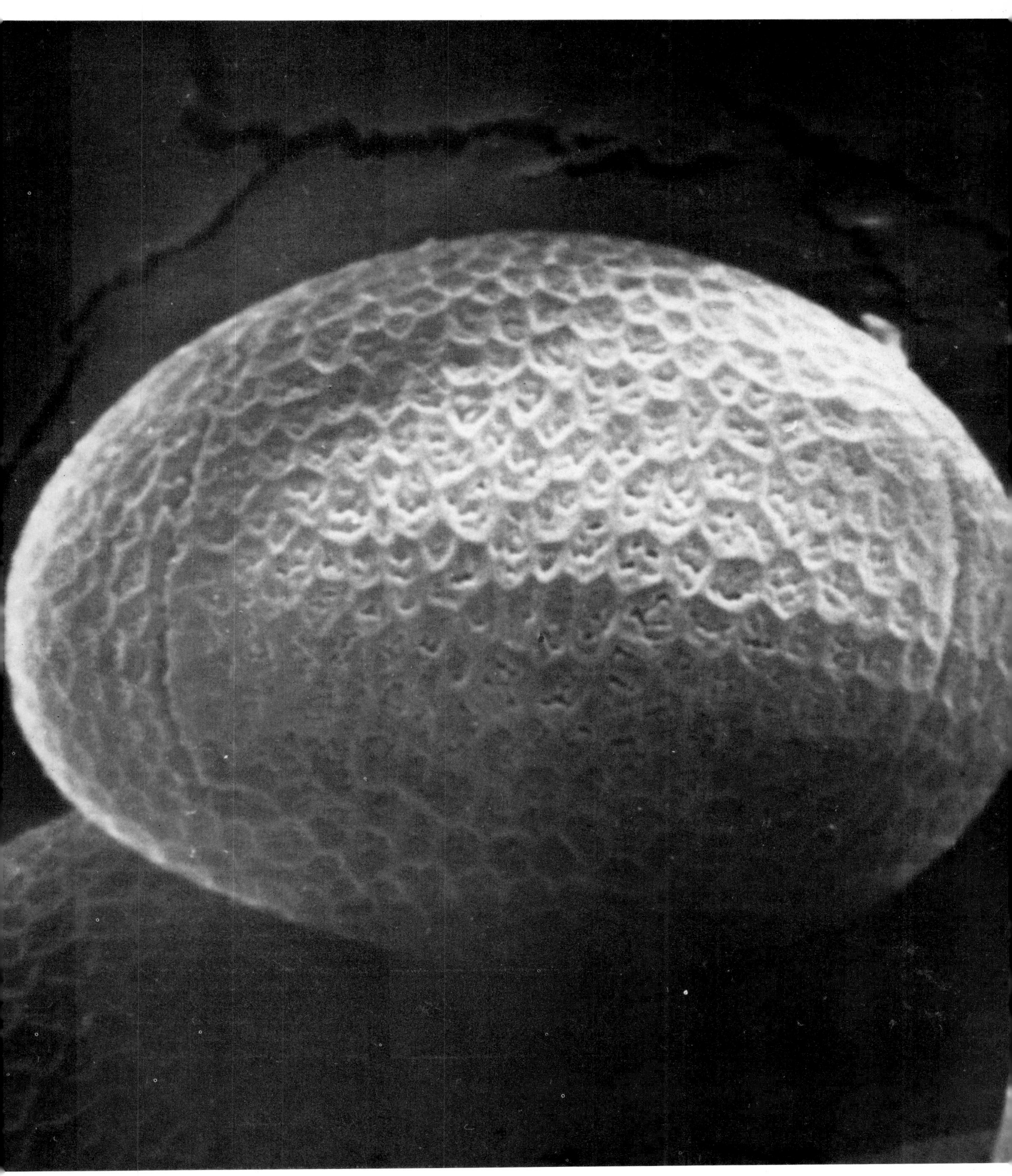

Wormwood pollen
6800 X—Dr. Patrick Echlin, University of Cambridge, England

Fat men pollen ►
7000 X—Dr. Patrick Echlin, University of Cambridge, England

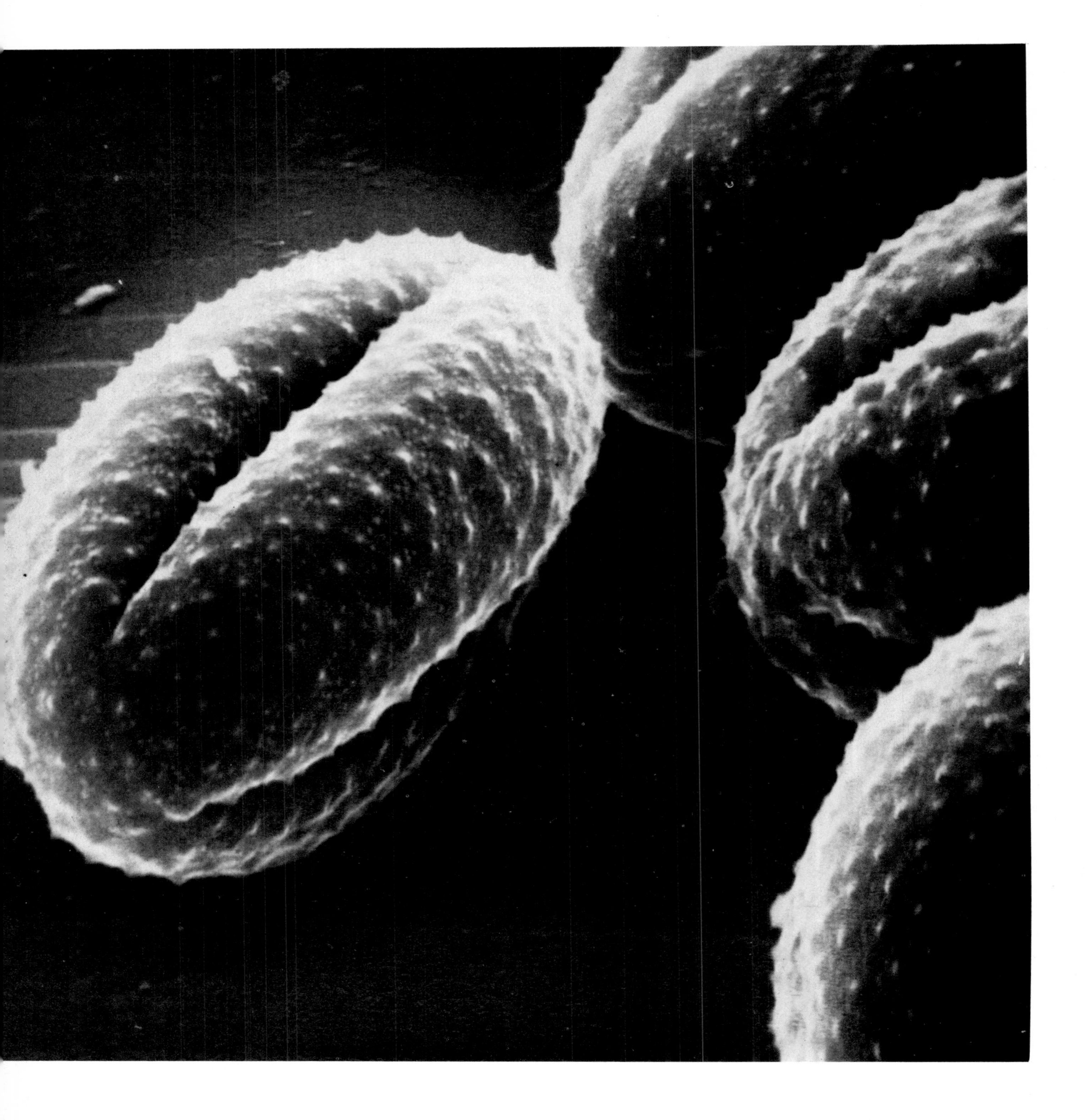

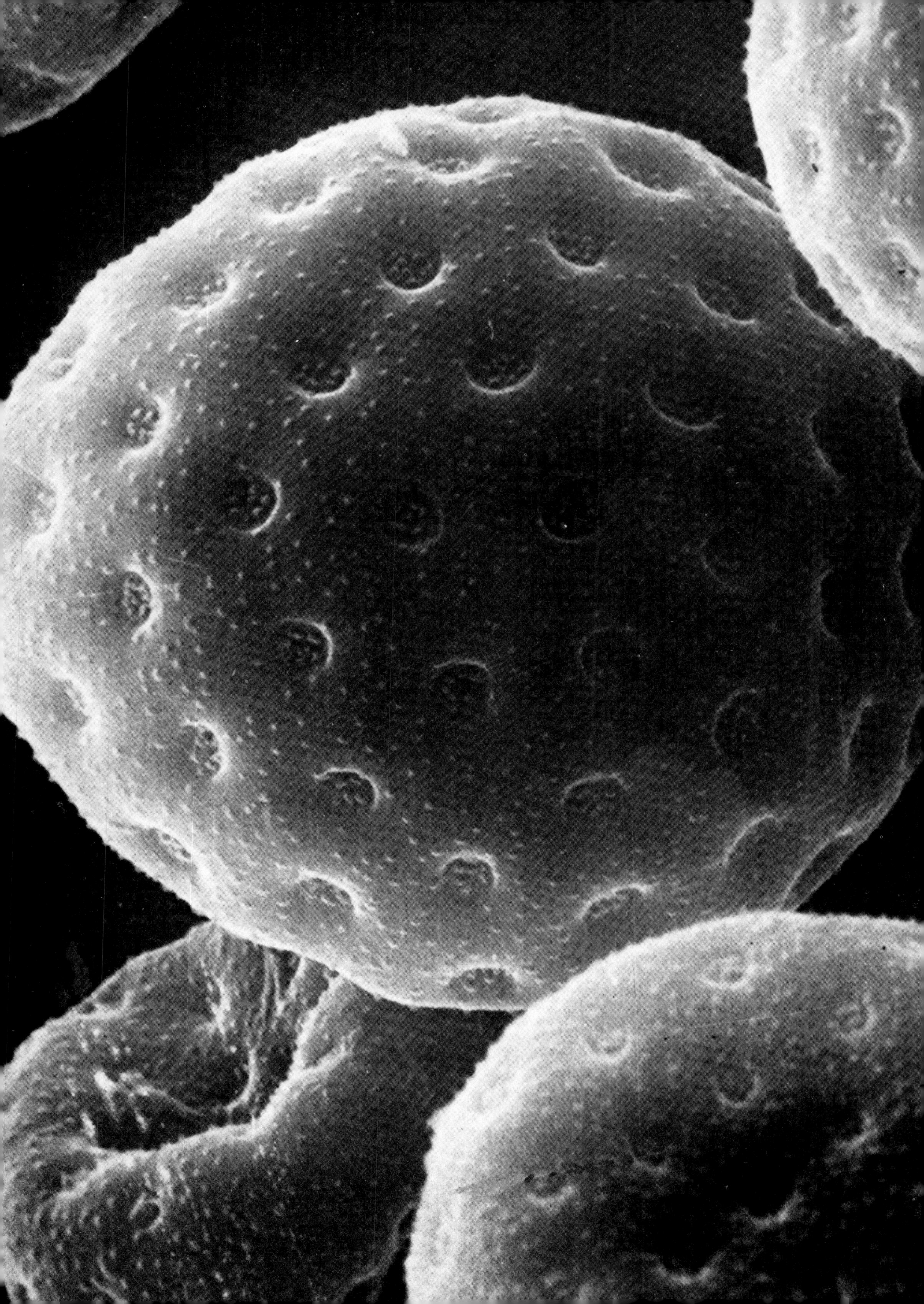

Soil Protozoa

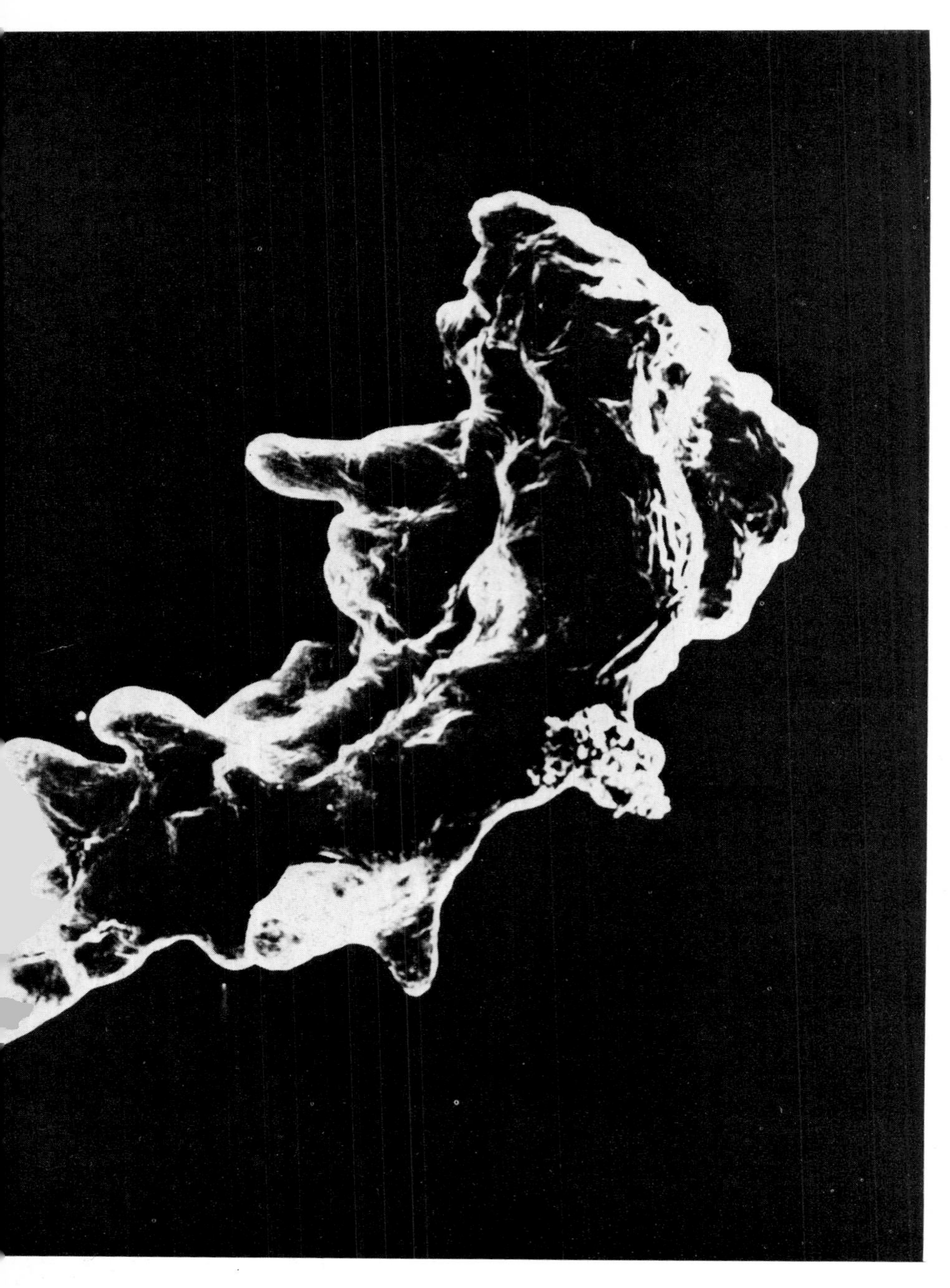

Protozoa are minute animals that live in the sea or stagnant fresh water, in the soil, and in the bodies of man and other animals—wherever there is moisture. Their most outstanding characteristic is that they are composed of a single cell. Thus they are the simplest—but by no means simple—animals, as these pictures show. Most protozoa are invisible to the unaided eye, some are as small as one micron—1/1000th of a millimeter. The creatures shown here range in size from *Amoeba proteus,* which is about 1/3rd of a millimeter long (barely visible under good light) to *Didinium*, which is 1/4th that size.

Because of their size, protozoa were not discovered until the invention of the light microscope in the 17th century. In 1647, the Dutch microscopist, Antony van Leeuwenhoek, focused his primitive instrument on a drop of pond water and was astonished by the teeming life he found there. He called the creatures "animalcules." The small size, stupendous numbers, and enormous powers of multiplication of the newly discovered class were used as evidence for the theory of spontaneous generation of life then popular. But in 1718, Louis Joblot retired that quaint notion by showing that nothing would generate spontaneously in water that was boiled and placed in a sealed container.

Protozoa exist in an unimaginable variety of sizes and types. More than 20,000 kinds have been identified, but there are undoubtedly many more. A few of them cause human diseases such as malaria, amoebic dysentery, and sleeping sickness.

This is *Amoeba proteus.* The lumpy looking protrusions are pseudopods, or false feet, temporary outflowings of protoplasm. An amoeba keeps putting out new psuedopods as he ambles along, and is, consequently, constantly changing his shape. He eats by surrounding a food particle and enclosing it within himself. Inside, enzymes digest the food and release the nutrients for the creature's use.
385 X—Dr. Eugene B. Small, University of Maryland

Here's a hirsute pair. *Nyctotherus ovalis*, the pear-shaped one, was found in the hind gut of a cockroach. The long one is the familiar *Paramecium multimicronucleatum*. It's easy to see why early investigators called it "the slipper animalcule." You may have watched this creature through the microscope in Zoology I as it navigated its way endlessly across a drop of water. Because it's easy to obtain and relatively large—and thus easy to see under a low-powered microscope—*Paramecium* is frequently used for basic teaching and simple studies.
Both of these creatures have what zoologists call holotrichous ciliary patterns. That means they're hairy all over. By beating the cilia rhythmically back and forth in a wavelike pattern, they propel themselves and waft food (chiefly bacteria) into their mouths. *Paramecium's* cilia tend to beat in a spiral pattern, so he rotates as he moves, screwing himself through the water like a corkscrew. ►
2300 X (top); 1350 X (bottom)—Dr. Eugene B. Small, University of Maryland

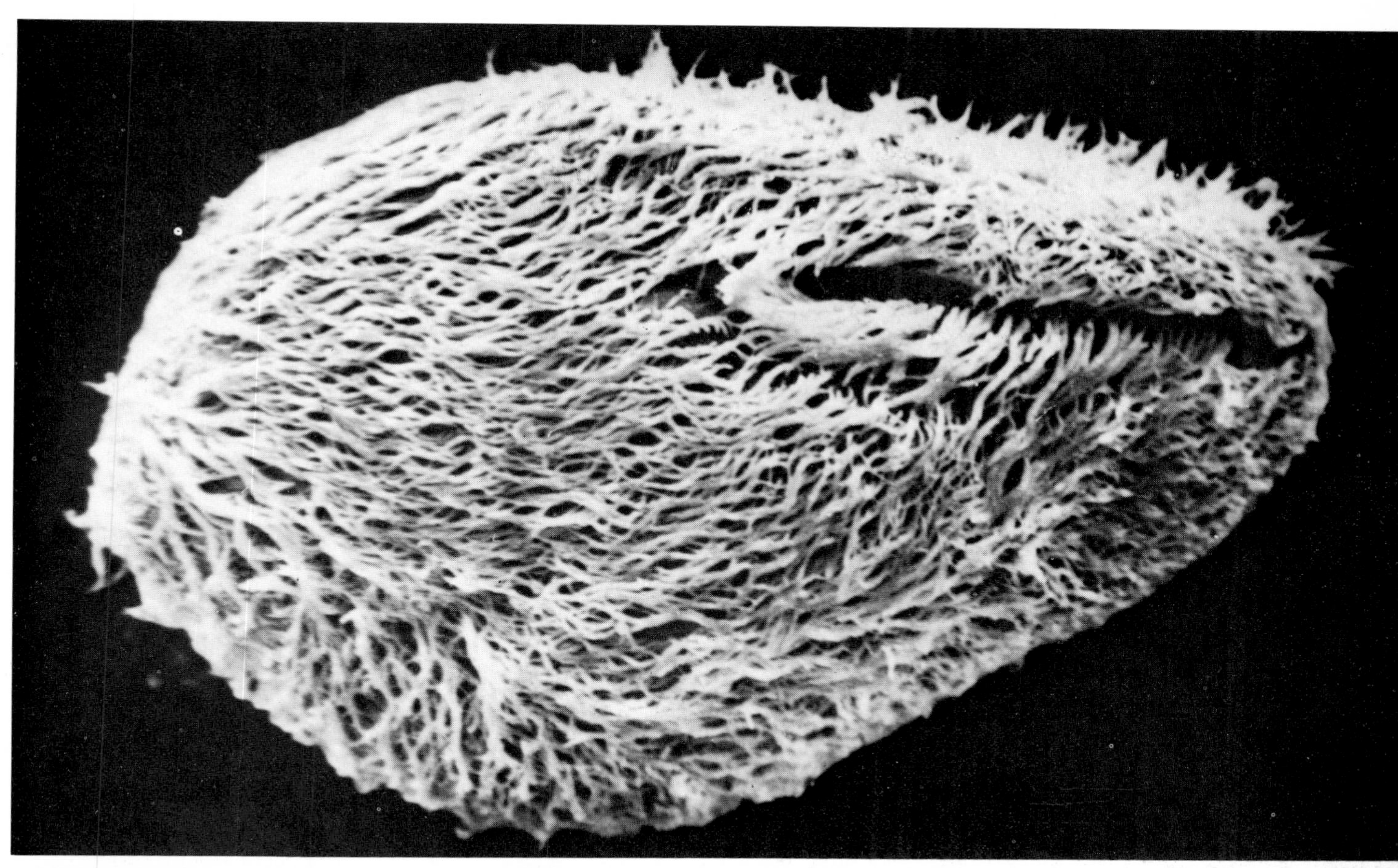

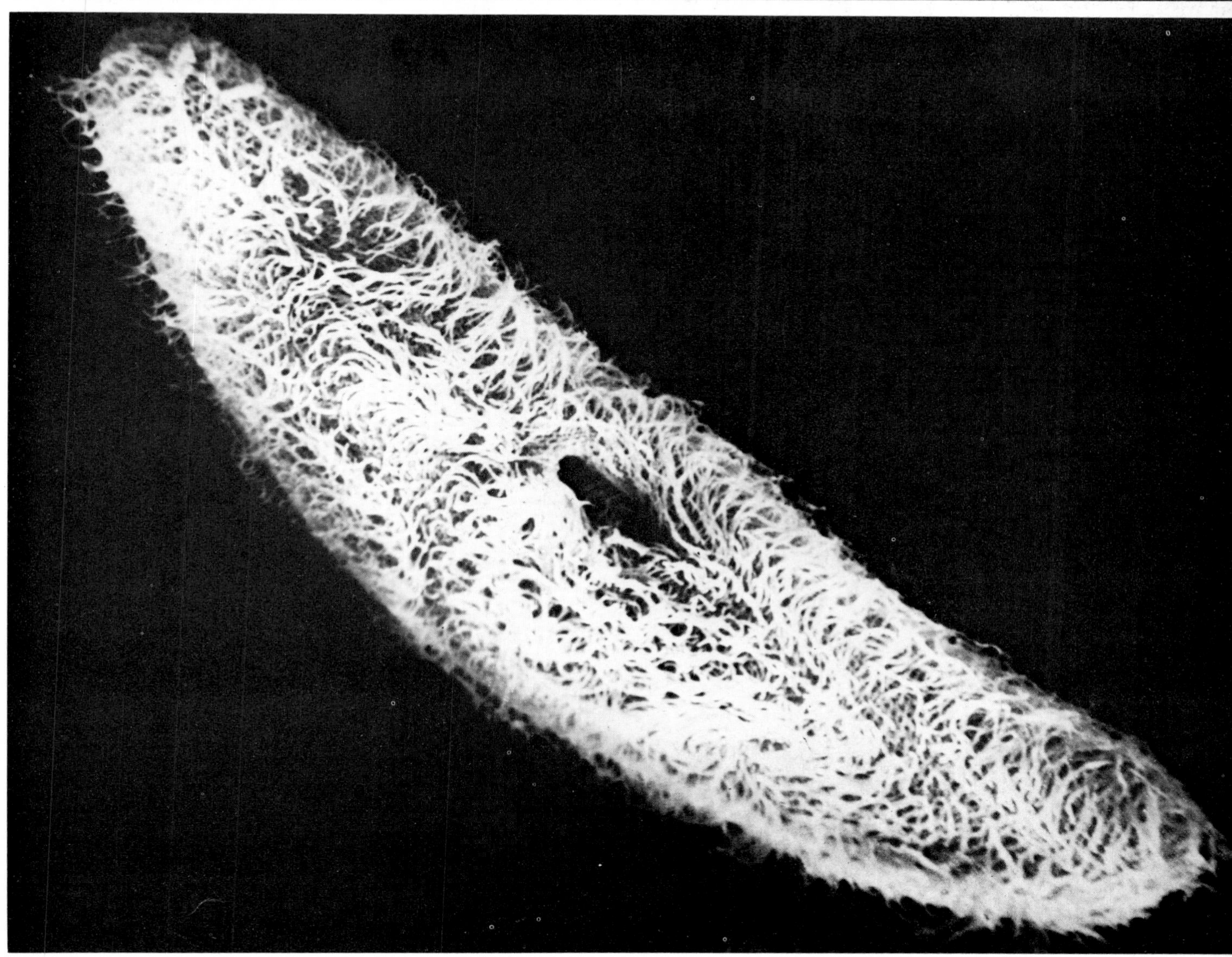

This *Paramecium* waved his cilia indiscreetly and propelled himself into becoming dinner for *Didinium nasutus.* In *Didinium's* cytopharynx (mouth), a special enzyme has already started to lyze or melt the cilia off the *Paramecium.*
1600 X—Dr. Eugene B. Small, University of Maryland

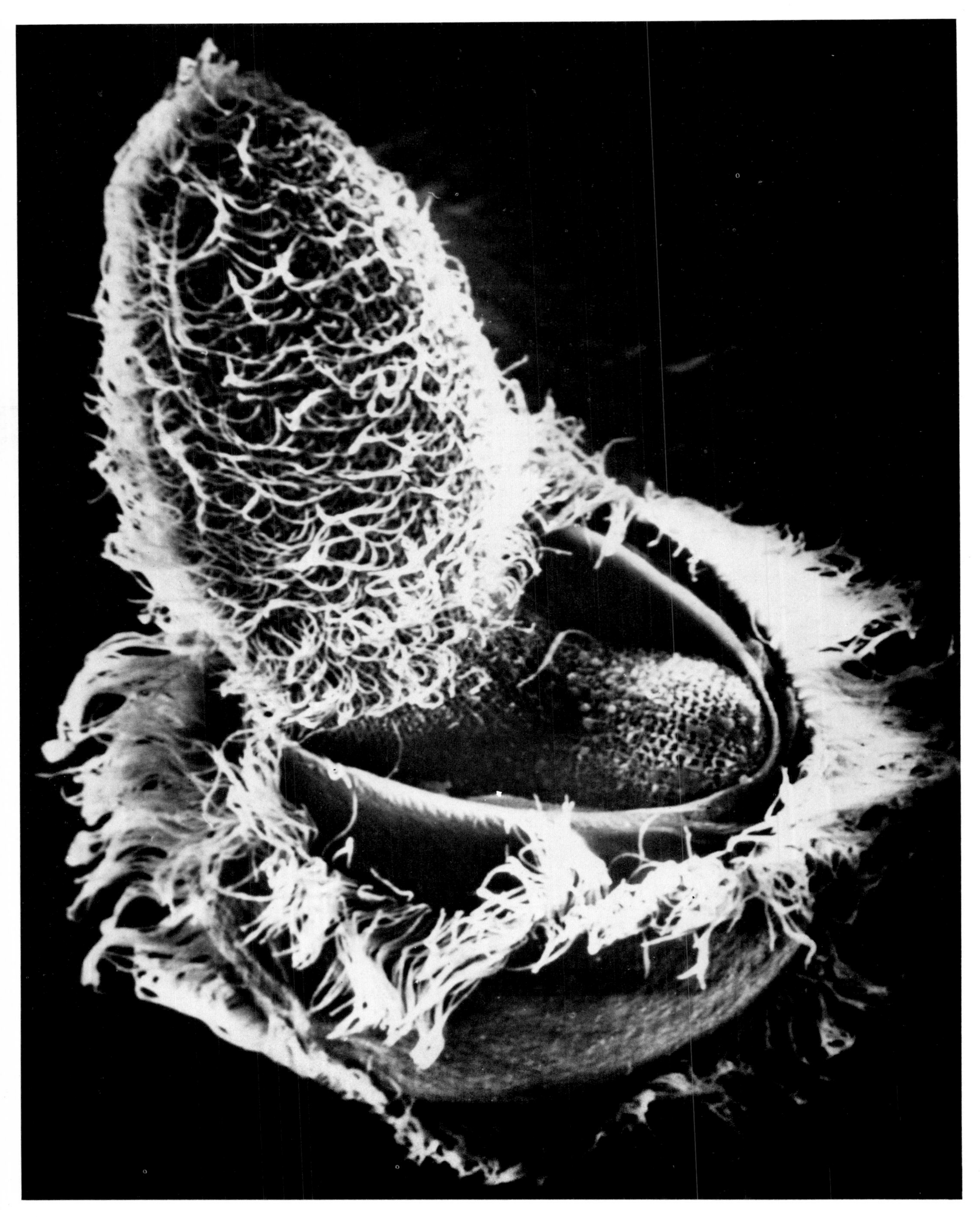

510

Photograph Credits

p. 7
Drawings by Jerome Kühl, Copyright 1970, Time Inc., Time–Life Books

pp. 12, 13, 139, 140, 141, 142
Professor J. Heslop-Harrison,
Dr. Y. Heslop-Harrison
Laboratory of Cytochemistry, Royal Botanic Gardens, Kew, England

pp. 18, 19 (top left), 20, 21, 23, 26, 28, 29, 56, 57, 60, 61, 65, 88, 100 (center and bottom), 102
Eastman Kodak Company, Kodak Park Division, Industrial Laboratory, Rochester, New York.
Photos by Mrs. Marianna Wilke, Mr. Fred Behnke.
Ant specimens provided by Dr. William Creighton, retired professor of zoology, City College, New York

pp. 62 (bottom), 86, 90, 113, 126
Dr. Emil Bernstein,
Eila Kairinen
Gillette Research Institute, Rockville, Md.

pp. 55 (bottom), 127
Dr. Emil Bernstein
Gillette Research Institute, Rockville, Md.

p. 105 (top)
Oak Ridge National Laboratory. Operated by Union Carbide Corporation for The U.S. Atomic Energy Commission

pp. 77 (top), 78 (bottom)
I. Kaufman Arenberg, M.D.
Department of Otolaryngology, Washington University School of Medicine, St. Louis, Missouri.
Supported by Grants USPHS-NIH-NSO-TI-5190 and Health Science Advancement Award F-304-FRO-6115

pp. 112, 114, 115, 116
Dr. Patricia N. Farnsworth
Barnard College.
Present address: College of Medicine and Dentistry of New Jersey, Newark, New Jersey

pp. 74, 75
Dr. Edwin R. Lewis,
Mrs. Paula K. Nemanic
Z. Zellforschung 123:441–457 (1972)

pp. 105 (bottom), 118
Irene Rodgers
Philips Electronic Instruments.
Philips EM 300 SEM photo

pp. 30, 31
Annemarie C. Reimschuessel,
John M. Kolyer
Chemical Research Center, Allied Chemical Corporation, Morristown, New Jersey

pp. 147–155
Dr. Patrick Echlin
The Botany School, University of Cambridge, England.
Copyright: *Scientific American*, April 1968

pp. 132–137
R. P. George, R. M. Albrecht, K. B. Roper
Department of Bacteriology, University of Wisconsin;
I. B. Sachs
Forest Products Laboratory, Madison, Wisconsin;
A. P. Mackenzie
Cryobiology Research Institute, Madison, Wisconsin

pp. 64, 71, 101 (bottom), 104, 117, 119
Richard Turnage
Advanced Metals Research Corp.
Pictures obtained with AMR Model 900 High Resolution Scanning Electron Microscope

pp. 24, 25
Dr. Don M. Rees, investigator
Microscopy by Division of Material Science and Engineering, University of Utah

pp. 124, 125
Dr. T. R. G. Gray
University of Liverpool.
124 and 125 (bottom): Copyright: Gray, T. R. G., *Stereoscan Electron Microscopy of Soil Microorganisms*; 125 (top): Copyright: Gray, T. R. G., and Williams, S. T., *Soil Microorganisms*, Oliver and Boyd, London

p. 82
G. A. Horridge
Australian National University, Canberra, Australia;
S. L. Tamm
Indiana University, Bloomington, Indiana

p. 98
Dr. R. J. Gerdes
present address: Scanatlanta Research Corp., 1645 Tully Circle N.E., Atlanta, Georgia 30329.
Composite grown by Dr. A. T. Chapman
Georgia Institute of Technology

pp. 8 (left), 73 (bottom)
Dr. Göran Bredberg
Akademiska Sjukhuset, Uppsala, Sweden
Copyright: Bredberg, Ades, and Engström; 1972 *Acta Otolaryng*, Suppl. 301

p. 35
Dr. G. W. Wharton, provided by the Acarology Laboratory, The Ohio State University, with the assistance of the Department of Otolaryngology and Dr. Arnold Brody.
Financial assistance provided by training grant AI-00216, National Institute of Allergy and Infectious Diseases, NIH

pp. 40–43
Dr. Richard H. Benson, Curator
Smithsonian Institution, Department of Paleobiology, United States
National Museum, Washington, D.C.

pp. 44, 45
Esso Production Research Company, Houston, Texas

p. 129 (bottom); Copyright Garrod and O'Grady (*Antibiotic and Chemotherapy*, Third Edition).
Dr. D. Greenwood
St. Bartholomew's Hospital, London

The following photographs were taken on the Cambridge "Stereoscan" Scanning Electron Microscope and reprinted from the Third Annual Stereoscan Colloquium sponsored by Kent Cambridge Scientific, Inc.:
pp. 8 (upper right), 33 (top), 34, 40–43, 77 (top), 78 (bottom), 87 (bottom), 98 (right), 99 (upper left), 120, 121, 132–137

The following photographs appeared in *Science* magazine:

pp. 12, 13 *Science*, vol. 167, cover, pp. 172–174
9 January, 1970 (Heslop-Harrison)

p. 32 *Science*, vol. 167, cover, pp. 1382–1383
6 March, 1970 (Wharton)

pp. 74, 75 *Science*, vol. 174, pp. 416–419
22 October, 1971 (Hillman and Lewis)

pp. 72, 73 *Science*, vol. 170, pp. 861–863
20 November, 1970 (Bredberg)

p. 82 *Science*, vol. 163, cover, pp. 817–818
21 February, 1969 (Horridge and Tamm)

p. 90 *Science*, vol. 173, cover
16 July, 1971 (Bernstein, Kairinen)

p. 108 *Science*, vol. 165, pp. 1140–1143
12 September, 1969 (Lewis, Zeevi, and Everhart)

p. 113 *Science*, vol. 173, cover
27 August, 1971 (Bernstein, Kairinen)

pp. 124, 125 *Science*, vol. 155, pp. 1668–1670
31 March, 1967 (Gray)

pp. 128, 129 *Science*, vol. 163, pp. 1076–1078
7 March, 1969 (Greenwood and O'Grady)